Date Due

Nilpotent rings

Notes on mathematics and its applications

General editors: Jacob T. Schwartz, *Courant Institute of Mathematical Sciences,* and Maurice Lévy, *Université de Paris*

E. Artin ALGEBRAIC NUMBERS AND ALGEBRAIC FUNCTIONS
R. P. Boas COLLECTED WORKS OF HIDEHIKO YAMABE
R. A. Bonic LINEAR FUNCTIONAL ANALYSIS
R. B. Burckel WEAKLY ALMOST PERIODIC FUNCTIONS ON SEMIGROUPS
M. Davis A FIRST COURSE IN FUNCTIONAL ANALYSIS
M. Davis LECTURES ON MODERN MATHEMATICS
J. Eells, Jr. SINGULARITIES OF SMOOTH MAPS
K. O. Friedrichs ADVANCED ORDINARY DIFFERENTIAL EQUATIONS
K. O. Friedrichs SPECIAL TOPICS IN FLUID DYNAMICS
A. Guichardet SPECIAL TOPICS IN TOPOLOGICAL ALGEBRAS
M. Hausner and J. T. Schwartz LIE GROUPS; LIE ALGEBRAS
P. Hilton HOMOTOPY THEORY AND DUALITY
S. Y. Husseini THE TOPOLOGY OF CLASSICAL GROUPS AND RELATED
 TOPICS
F. John LECTURES ON ADVANCED NUMERICAL ANALYSIS
A. M. Krall STABILITY TECHNIQUES FOR CONTINUOUS LINEAR
 SYSTEMS
R. L. Kruse and D. T. Price NILPOTENT RINGS
P. Lelong PLURISUBHARMONIC FUNCTIONS AND POSITIVE
 DIFFERENTIAL FORMS
H. Mullish AN INTRODUCTION TO COMPUTER PROGRAMMING
F. Rellich PERTURBATION THEORY OF EIGENVALUE PROBLEMS
J. T. Schwartz DIFFERENTIAL GEOMETRY AND TOPOLOGY
J. T. Schwartz NONLINEAR FUNCTIONAL ANALYSIS
J. T. Schwartz W-* ALGEBRAS
G. Sorani AN INTRODUCTION TO REAL AND COMPLEX MANIFOLDS
J. L. Soulé LINEAR OPERATORS IN HILBERT SPACE
J. J. Stoker NONLINEAR ELASTICITY

Additional volumes in preparation

Nilpotent rings

Robert L. KRUSE
Mathematics Department, Sandia Laboratory
New Mexico

and

David T. PRICE
Wheaton College
Illinois

GORDON AND BREACH
Science Publishers
NEW YORK   LONDON   PARIS

Copyright © 1969 by GORDON AND BREACH, SCIENCE PUBLISHERS, INC.
150 Fifth Avenue, New York, N.Y. 10011

Library of Congress catalog card number: 78–84937

Editorial office for the United Kingdom:
Gordon and Breach, Science Publishers Ltd.
12 Bloomsbury Way
London W.C.1

Editorial office for France:
Gordon & Breach
7–9 rue Emile Dubois
Paris 14e

Distributed in Canada by:
The Ryerson Press
299 Queen Street West
Toronto 2B, Ontario

Preface

THE STRUCTURE THEORY of non-commutative rings falls naturally into three parts: the study of semi-simple rings; the study of radical rings; and the construction of rings with a given radical and semi-simple factor ring. The first of these problems has been handled by far the most successfully. This book is intended to add to the understanding of the second problem by presenting for the first time a unified treatment of most of the significant known results about nilpotent rings and algebras. We hope to provide a reference work for researchers in the field, as well as an introduction to the subject for those who wish to explore it. Most of the results in this book have not previously appeared in the literature, while a few have had an unhappy history of republication of special cases.

The early results of the book suggest an analogy between the theory of nilpotent rings and the theory of nilpotent groups, and many of the later results are formulated by means of this analogy. In particular, we owe a great debt to the fundamental discoveries of Philip Hall concerning p-groups [1, 3]. The ring results, however, usually possess direct proofs which, because of the added structure available in a ring, are often simpler than the proofs of the corresponding group results.

The main topic of this book is the structure of nilpotent rings, and we have generally omitted the many interesting, related topics, such as conditions implying nilpotence, the study of other classes of radical rings (e.g. locally nilpotent, nil rings), and the structure of commutative nilpotent rings. The 1961 notes of Herstein [1] present several important results about locally nilpotent and nil rings. Finite-dimensional commutative nilpotent algebras, regarded as algebras of matrices, are studied in the book of Suprunenko and Tyshkevich [1]. Embedding theorems for linear spaces of nilpotent matrices are obtained in a series of papers by Gerstenhaber [1–5]. The various possible definitions of a radical and their relations to the structure of a ring are presented in the book of Divinsky [2]. Cohomology theory as developed, for example, by D. K. Harrison, has interesting applications to commutative nilpotent algebras.

We have attempted to keep the proofs in the present book as elementary and self-contained as possible, but for background and

v

motivation the reader should have some acquaintance with non-commutative ring theory up through the Wedderburn–Artin structure theory for semi-simple rings, and with the elementary theory of finite p-groups, as may be found in any of the standard textbooks. The ring theory book of McCoy [2] is particularly recommended.

Chapter 1 and parts of Chapter 2 present material basic to the entire book, but Chapters 3 through 5 are mutually independent and may be read in any order. The appendix, Chapter 6, is included for reference purposes. In addition to the works cited in the text, the bibliography includes all papers available to the authors on the structure of nilpotent rings.

The interested reader may note that many of the results of this book hold without the assumption of associativity. In particular, almost all of the results formulated by means of the analogy with p-groups will hold also for nilpotent Lie rings.

The authors wish to express their appreciation to the Sandia Laboratory and the United States Atomic Energy Commission for sponsoring much of the research presented here, and for providing unfailing support for the preparation of this book. The first author was associated with the Sandia Laboratory throughout the preparation of the book, and the second author during the summers of 1966 and 1967.

It is a pleasure to acknowledge our indebtedness to the secretarial staff of the Sandia Laboratory, and especially to Mrs. Josephine Emery, who not only typed several versions of the entire manuscript, but showed careful accuracy in helping to organize and check the bibliography and index. Finally, we wish to thank the officers and staff of Gordon and Breach for their carefulness and efficiency throughout the production of this book.

January 1969   R. L. KRUSE
                                                D. T. PRICE

Contents

Preface   v

CHAPTER I   **Elementary results** **1**

    1 Definitions and notation   1
    2 Annihilators   2
    3 Annihilator series   4
    4 Subdirectly irreducible nilpotent rings   7
    5 The primary decomposition   9
    6 The circle group   10

CHAPTER II   **Examples of nilpotent rings** **17**

    1 Representations and rings of matrices   17
    2 Free rings   19
    3 The construction of rings by generators and
      relations   20
    4 Modular group algebras of p-groups   24

CHAPTER III   **Families and capability** **27**

    1 The family classification   27
    2 On the rings in a family   29
    3 An example: the nilpotent rings of order p^3 34
    4 Nilpotent rings with chain conditions   40
    5 Capability; the construction of families   42
    6 Conditions for capability   48

CHAPTER IV   **The subring structure of nilpotent rings**   .   .   **51**

    1 On generating rings   51
    2 Automorphisms   54
    3 Enumeration results   59
    4 Nilpotent rings with only one subring of a
      given order   62

5 Nilpotent p-rings with one subring of
  order p 67
6 Rings in which all subrings are ideals   68

CHAPTER V   **Counting finite nilpotent rings** **80**

1 Introduction   80
2 Asymptotic results   81
3 Asymptotic results for non-nilpotent rings   90

CHAPTER VI   APPENDIX **The construction of some special rings**   **95**

1 Some nilpotent p-rings of rank 2   95
2 The nilpotent algebras of dimensions 4   99
3 Assume dim $A = 4$, $A^3 \neq 0$ 101
4 Assume dim $A = 4$, dim $A^2 = 1$, $A^3 = 0$,
  and A is not reducible   102
5 Assume dim $A = 4$, dim $A^2 = 2$, $A^3 = 0$ 104

Bibliography   109

List of special notation   123

Index   125

CHAPTER I

Elementary results

1 Definitions and notation

A RING R is called *nilpotent* if there is a natural number n such that all products of n elements from R are 0. The smallest such n is called the *exponent* of R and is denoted exp R. If $R^2 = 0$, the ring R is called *null*. The following observation is of frequent, implicit use in the study of nilpotent rings.

1.1.1 *A subring S and a homomorphic image Q of a nilpotent ring R are nilpotent, and* exp $S \leq$ exp R, exp $Q \leq$ exp R. *If, on the other hand, I is a nilpotent two-sided ideal of a ring R and R/I is nilpotent, then R is nilpotent and* exp $R \leq$ exp $I \cdot$ exp (R/I).

An element of a ring is called *nilpotent* if some power of the element is 0. A ring in which every element is nilpotent is called *nil*. Easy examples show that a nil ring need not be nilpotent.

If S is a subset of ring R, the smallest subring of R which contains all elements of S (the *subring generated* by S) is denoted $\langle S \rangle$. We shall write $S + T$ for $\langle S \cup T \rangle$ and ST for $\langle st \mid s \in S, t \in T \rangle$. The symbol $\oplus$ denotes a direct sum. In the infinite case only restricted direct sums are considered. An integer may be interpreted as an operator on a ring R, even when R does not contain an identity. For example, we write $(1 + x)R$ to mean $\{r + xr \mid r \in R\}$. The symbol $\cong$ denotes isomorphism; the symbol $-$ between subsets of a ring denotes set difference.

The *characteristic* of a non-empty subset S of a ring, denoted char S is the smallest natural number n such that $nx = 0$, all $x \in S$, or 0 if no such number exists. The additive group of a ring R is denoted R^+. The *order* of a finite ring R, denoted $\mid R \mid$, is the order of R^+. The *index* of a subring S in a ring R, denoted $[R:S]$, is the index of S^+ in R^+. Similarly, in a ring R the terms *height, span, rank, basis, type, cyclic*, and *quasi-cyclic* are defined according to their meanings for the additive group R^+. For the definitions of abelian group terms, see Fuchs [2] or Kaplansky [1]. The symbol $C(p^i)$ will denote the cyclic group of order p^i, if i is a natural number, and the Prüfer quasi-cyclic p-group if $i = \infty$. The type of an abelian p-group will be denoted $_p(n_1, \ldots, n_k)$. The prime p may be omitted when its value is clear from the context.

A ring A is called an *algebra* over a field F if A is a vector space over F and if $\alpha \in F$, $a, b \in A$ implies $\alpha(ab) = (\alpha a)b = a(\alpha b)$. The multiplicative group of non-zero elements in the field F is denoted F^*. The term q-*algebra* denotes an algebra over $GF(q)$, the field of q elements, where q is a prime power. In the case of a prime p we shall identify $GF(p)$ with the ring of rational integers mod p. For each prime p a ring R has two naturally associated p-algebras, which are sometimes useful in studying the structure of R. The first of these is R/pR; the second is the ideal $R_p = \{x \in R \mid px = 0\}$.

Essentially all the results stated for rings in this book will also hold for algebras, where all subrings are replaced by the corresponding subalgebras, and either finiteness or finite rank is replaced by finite dimensionality. Note that in the case when the field is infinite, then the rank of the algebra as a ring is infinite, but if the algebra is finite dimensional over the field then the analogues of results proved for rings of finite rank will still hold. In the other direction, however, it is clear that a result on algebras need have no analogue which holds for other classes of rings. In proving a result for algebras we shall frequently refer to a preceding ring result, when the analogue for algebras is in fact required. This ambiguity should cause no difficulty for the reader.

Unless otherwise specified, all rings and algebras considered in this book will be assumed to be associative. The unqualified word "ideal" always means two-sided ideal. Many results are stated only for the right side. Dual results, of course, hold for the left side.

2 Annihilators

The *left, right*, and *two-sided annihilators* of a non-empty subset S of a ring R are defined, respectively, to be

$$\mathfrak{A}_l(S) = \{x \in R \mid xS = 0\}.$$
$$\mathfrak{A}_r(S) = \{x \in R \mid Sx = 0\}.$$
$$\mathfrak{A}(S) = \{x \in R \mid xS = Sx = 0\}.$$

The unqualified word "annihilator" means two-sided annihilator. There are several immediate observations on annihilators.

1.2.1 *Let S and T be non-empty subsets of a ring R. Then*
(1) *If $S \subseteq T$, then $\mathfrak{A}_r(T) \subseteq \mathfrak{A}_r(S)$ and $\mathfrak{A}(T) \subseteq \mathfrak{A}(S)$.*
(2) *$\mathfrak{A}_r(S)$ is a right ideal of R, and $\mathfrak{A}(S)$ is a subring of R.*
(3) *If S is a right ideal of R, then $\mathfrak{A}_r(S)$ is a two-sided ideal of R. If S is an ideal of R, so is $\mathfrak{A}(S)$.*
(4) *$\mathfrak{A}_l(\mathfrak{A}_r(S)) \geq S$ and $\mathfrak{A}(\mathfrak{A}(S)) \geq S$.*

(5) $\mathfrak{A}_r(\mathfrak{A}_l(\mathfrak{A}_r(S))) = \mathfrak{A}_r(S)$ and $\mathfrak{A}(\mathfrak{A}(\mathfrak{A}(S))) = \mathfrak{A}(S)$.

(6) $\mathfrak{A}_r(\mathfrak{A}_l(S)) = S$ if and only if $S = \mathfrak{A}_r(T)$ for some subset T of R. $\mathfrak{A}(\mathfrak{A}(S)) = S$ if and only if $S = \mathfrak{A}(T)$ for some subset T of R.

(7) Every subgroup of $(\mathfrak{A}_r(S))^+ \cap \langle S \rangle^+$ is a left ideal in $\langle S \rangle$. Every subgroup of $(\mathfrak{A}(S))^+ \cap \langle S \rangle^+$ is a two-sided ideal in $\langle S \rangle$.

(8) (Hinohara [1]) For $i = 1, \ldots, n$ let X_i and Y_i be non-empty subsets of R such that $\mathfrak{A}_r(X_i) \subseteq \mathfrak{A}_l(Y_i)$. Then $\mathfrak{A}_r(X_1 \ldots X_n) \subseteq \mathfrak{A}_l(Y_1 \ldots Y_n)$.

1.2.2 Let S and T be non-empty, non-zero subsets of a nilpotent ring. Let $P = \{st \mid s \in S, t \in T\}$. Then $T \nsubseteq P$.

1.2.3 Let $S \neq 0$ be a subring of a nilpotent ring R. Then

(1) If S is a right ideal of R, then $S \cap \mathfrak{A}_l(R) \neq 0$.

(2) If S is an ideal of R, then $S \cap \mathfrak{A}(R) \neq 0$.

1.2.4 (Frobenius [2]) If $xy = x$ for elements x and y of a nilpotent ring, then $x = 0$. If $x_1, x_2, \ldots, x_n$ are elements of a nilpotent algebra such that $x_1 x_2 \ldots x_n \neq 0$, then the products

$$x_1, x_1 x_2, \ldots, x_1 x_2 \ldots x_n$$

are all linearly independent.

All of these results follow directly from the definitions. As an example we shall prove **1.2.2**. Assume that the hypotheses hold, but $T \subseteq P$. Choose $t \in T$. From $T \subseteq P$ follows $t = s_1 t_1$ for some $s_1 \in S$, $t_1 \in T$. Now suppose inductively that for some $k \geq 1$, $t = s_1 \ldots s_k t_k$ for suitable $s_i \in S$, $1 \leq i \leq k$, $t_k \in T$. Again, since $T \subseteq P$, $t_k = s_{k+1} t_{k+1}$ for some $s_{k+1} \in S$, $t_{k+1} \in T$. Thus $t = s_1 \ldots s_{k+1} t_{k+1}$. Since S is contained in a nilpotent ring, there is a natural number n such that $s_1 \ldots s_n = 0$. Hence $t = 0$, so $T = 0$, contrary to hypothesis.

We conclude this section by looking at the relations between the characteristic of a ring and that of its annihilator.

1.2.5 If R is any ring, then char $R/\mathfrak{A}(R) = $ char R^2.

Proof Suppose char $R^2 = n \neq 0$. Then, for all x, $y \in R$, $(nx)y = 0$ and $n(yx) = y(nx) = 0$, so $nx \in \mathfrak{A}(R)$. Thus $m = $ char $R/\mathfrak{A}(R)$ divides n. On the other hand, from $mx \in \mathfrak{A}(R)$, all $x \in R$, follows $mxy = 0$, all x, $y \in R$, so n divides m. Thus $m = n$. A similar argument shows that char $R^2 = 0$ if and only if char $R/\mathfrak{A}(R) = 0$.

1.2.6 Let R be a nilpotent ring. Then char $R = 0$ if and only if char $\mathfrak{A}(R) = 0$.

Proof If char $\mathfrak{A}(R) = 0$, then certainly char $R = 0$. Let $e = \exp R$, and suppose char $R = 0$, but $m = $ char $\mathfrak{A}(R) \neq 0$. Since $R^{e-1} \subseteq \mathfrak{A}(R)$, there is a smallest natural number i such that $n = $ char $R^i \neq 0$. Since i is minimal, there is some $x \in R^{i-1}$ such that $mnx \neq 0$. From $xR + Rx \subseteq R^i$ follows $n(xR) = n(Rx) = 0$, so $nx \in \mathfrak{A}(R)$. But $m\,\mathfrak{A}(R) = 0$, so $mnx = 0$, contradicting the choice of x. Thus if char $R = 0$, then char $\mathfrak{A}(R) = 0$.

3 Annihilator series

A chain of ideals of a ring R,

$$0 = I_k \subset I_{k-1} \subset \ldots \subset I_1 = R,$$

is called an *annihilator series* for R if $I_j/I_{j+1} \subseteq \mathfrak{A}(R/I_{j+1})$ for $j = 1, 2, \ldots, k - 1$. If R is a nilpotent ring of exponent n, then

$$0 = R^n \subset R^{n-1} \subset \ldots \subset R^1 = R$$

is an annihilator series for R. If, on the other hand,

$$0 = I_k \subset I_{k-1} \subset \ldots \subset I_1 = R$$

is an annihilator series for a ring R, then $R^j \subseteq I_j$, $j = 1, \ldots, k$, so that R is nilpotent, and exp $R \leq k$. Because $R^j \subseteq I_j$, we call the series of powers of a nilpotent ring R the *lower annihilator series* of R.

Next let us inductively define a series of ideals of a ring R by $A_0 = 0$ and A_{i+1} is the ideal which contains A_i such that $A_{i+1}/A_i = \mathfrak{A}(R/A_i)$ for $i = 0, 1, 2, \ldots$. If

$$0 = I_k \subseteq \ldots \subseteq I_1 = R$$

is any annihilator series for R, then $I_{k-j} \subseteq A_j$, $0 \leq j < k$. Because of this we call the series

$$0 = A_0 \subset A_1 \subset \ldots \subset A_{n-1} = R$$

the *upper annihilator* series of a nilpotent ring R of exponent n. For $i \geq n$ define $A_i = R$. We now summarize the above results. See figure 1.

1.3.1   THEOREM   *A ring is nilpotent if and only if it possesses an annihilator series. Let R be a nilpotent ring of exponent n and*

$$0 = I_k \subseteq \ldots \subseteq I_1 = R$$

an annihilator series of R. Then $k \geq n$ and, for $1 \leq j \leq k$, $R^j \subseteq I_j$ and $I_{k-j+1} \subseteq A_{j-1}$.

1.3.2   COROLLARY   *The upper and lower annihilator series of a nilpotent ring have the same length.*

Theorem **1.3.1** suggests an analogy between the theories of nilpotent groups and nilpotent rings. Many results on nilpotent rings may be correctly formulated by taking nilpotent group results and replacing commutation by ring product. The following correspondences result:

Groups	Rings
$[a, b]$	$a\dot{o}$
commute	annihilate
abelian	null
(upper, lower) central series	(upper, lower) annihilator series.

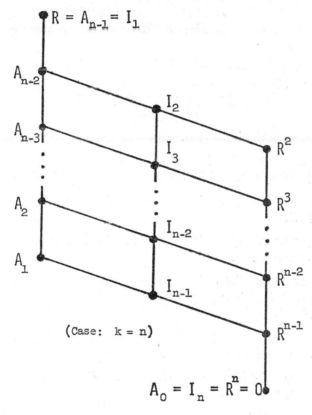

FIGURE 1

Since ring product is associative, proofs of corresponding results are often easier for rings than for groups. The next result is analogous to one obtained for groups by C. Sims [1].

1.3.3 THEOREM *Let S be a subring of a ring R. Suppose for some $i \geq 1$ that $R^i = S^i + R^{i+1}$. Then, for all $j \geq i$ and all $k \geq 1$, $R^j = S^j + R^{j+k}$.*

Proof We prove the result by induction on the pairs (k, j), ordered lexicographically. For a discussion of lexicographic ordering see, for example, van der Waerden [1]. Suppose $k = 1$. The result holds by hypothesis for $j = i$. Suppose $j > i$ and $R^{j-1} = S^{j-1} + R^j$. Then

$$
\begin{aligned}
R^j = R^{j-1}R &= (S^{j-1} + R^j)R \\
&\subseteq SR^{j-1} + R^{j+1} \\
&= S(S^{j-1} + R^j) + R^{j+1} \\
&\subseteq S^j + R^{j+1}.
\end{aligned}
$$

The reverse containment is obvious. Suppose $k > 1$ and the result is true for all pairs preceding (k, j). Then $R^j = S^j + R^{j+k-1}$. Since $(1, j + k - 1)$ precedes (k, j), $R^{j+k-1} = S^{j+k-1} + R^{j+k}$, so

$$
\begin{aligned}
R^j &= S^j + S^{j+k-1} + R^{j+k} \\
&= S^j + R^{j+k},
\end{aligned}
$$

which completes the induction.

1.3.4 COROLLARY *If S is a subring of a nilpotent ring R such that $R^i = S^i + R^{i+1}$ for some $i \geq 1$, then $R^j = S^j$ for all $j \geq i$.*

1.3.5 COROLLARY *If M is a maximal ideal of a nilpotent ring R, then $R^2 \subseteq M$.*

1.3.6 COROLLARY *If M is a maximal ideal of a nilpotent ring R, then $[R:M]$ is a prime.*

1.3.7 COROLLARY *Let S be a proper subring of a nilpotent ring R. Then S is contained in a proper ideal of R. Every maximal subring of a nilpotent ring is an ideal.*

Because of its frequent application, we state and prove separately the following special case of **1.3.4**.

1.3.8 LEMMA *Let S be a subring of a nilpotent ring R such that $R = S + R^2$. Then $R = S$.*

Proof Since R is nilpotent, there is a smallest natural number k with $R^k = S^k$. If $k > 1$, then

$$
R^{k-1} = (S + R^2)^{k-1} \subseteq S^{k-1} + R^k = S^{k-1}.
$$

Thus $k = 1$, q.e.d.

1.3.9 REFINEMENT THEOREM *Let S and T be subrings of a nilpotent* *ring R, with $S \subset T$. Then there are subrings of R*

$$S = S_1 \subseteq S_2 \subseteq \ldots \subseteq S_k = T,$$

where $k \leq \exp R$, and

(1) S_i *is an ideal in S_{i+1}, $1 \leq i \leq k - 1$.*

(2) S_{i+1}/S_i *is a null ring, $1 \leq i \leq k - 1$.*

(3) *If S and T are right (resp. two-sided) ideals of R, then for $1 < i \leq k$, the subrings S_i may be chosen as right (resp. two-sided) ideals of R for which $S_i R \subseteq S_{i-1}$ (resp. $S_i/S_{i-1} \subseteq \mathfrak{A} (R/S_{i-1})$).*

Proof In the subring case choose k minimal such that $T^k \subseteq S$, and define $S_i = S + T^{k-i+1}$. In the right (two-sided) ideal case choose $k = \exp R$ and define $S_i = S + T R^{k-i}$ $(S_i = S + \sum_{j=0}^{k-i} R^j T R^{k-i-j})$. The assertions of the theorem follow directly.

1.3.10 COROLLARY *If, for some i, $1 \leq i \leq k - 1$, $[S_{i+1}:S_i]$ is not a prime, then the chain of subrings may be further refined, with the preservation of (1), (2), and (3).*

1.3.11 COROLLARY *If R is a nilpotent ring of order p^n, then $\exp R \leq n + 1$.*

1.3.12 COROLLARY *If R is a nilpotent ring of order p^n, then R has an ideal of each possible order p^i, $0 \leq i \leq n$.*

4 Subdirectly irreducible nilpotent rings

A ring S is a *subdirect sum* of rings R_1 and R_2 if S is a subring of $R_1 \oplus R_2$ and the natural projections of S into R_1 and R_2 are epimorphisms. S is *subdirectly irreducible* if, whenever S is isomorphic to a subdirect sum, then the natural projection of S onto at least one of the summands is an isomorphism. Subdirect sums are related to ideals by the following well-known result, which may be found, for example, in McCoy [1, 2]:

1.4.1 LEMMA *Every representation of a ring R as a subdirect sum of rings S_i, i ranging over an index set I, corresponds to a set of ideals K_i, $i \in I$, for which $R/K_i \cong S_i$ and $\bigcap K_i = 0$. Every set of ideals K_i for which $\bigcap K_i = 0$, conversely, corresponds to a representation of R as a subdirect sum of the rings R/K_i.*

In this section we characterize the class of subdirectly irreducible nilpotent rings and discuss several examples. We need the following elementary result about abelian groups.

1.4.2 LEMMA *Let A be an abelian group for which the intersection of all non-zero subgroups is not zero. Then $A \cong C(p^i)$ for some prime p and some i, $1 \leq i \leq \infty$.*

1.4.3 THEOREM *Let R be a nilpotent ring. Then R is subdirectly irreducible if and only if $(\mathfrak{A}(R))^+ \cong C(p^i)$ for some prime p and some i, $1 \leq i \leq \infty$.*

Proof By **1.4.1** R is subdirectly irreducible if and only if the intersection of non-zero ideals of R is not zero. Since every additive subgroup of $\mathfrak{A}(R)$ is an ideal in R, and since by (2) of **1.2.3** every non-zero ideal of R has a non-zero intersection with $\mathfrak{A}(R)$, it follows that the intersection of the non-zero ideals of R equals the intersection of the non-zero subgroups of $(\mathfrak{A}(R))^+$, which, by Lemma **1.4.2**, is not zero if and only if $(\mathfrak{A}(R))^+ \cong C(p^i)$ for some prime p and some i, $1 \leq i \leq \infty$.

The null rings with additive groups $C(p^i)$ provide examples of subdirectly irreducible nilpotent rings with all possible annihilator structures. Note that the annihilator of a subdirectly irreducible nilpotent ring R can never contain elements of characteristic zero. The following example, however, shows that R itself may still contain such elements.

1.4.4 EXAMPLE *of a subdirectly irreducible nilpotent ring R which contains elements of characteristic* 0.

Let R^+ be the (restricted) direct sum of a quasi-cyclic p-group and a countable number of infinite cyclic groups with generators y_i, $i = 1, 2, \ldots$. Denote generators of the quasi-cyclic group by x_i, $i = 1, 2, \ldots$, so that $px_1 = 0$ and $px_i = x_{i-1}$, $i \geq 2$. Define multiplication in R by $x_i R = R x_i = 0$, $1 \leq i$, and $y_i y_j = x_{i \cdot j}$, $i, j = 1, 2, \ldots$, with extension by distributivity to all of R. All products of three elements from R are then 0, so associativity and nilpotence are trivial. A direct argument shows that $\mathfrak{A}(R) = \langle x_i \mid i = 1, 2, \ldots \rangle$.

1.4.5 THEOREM *Let ring S be a subdirect sum of rings R_1 and R_2. Let $S_i = S \cap R_i$, $i = 1, 2$. Then S_i is an ideal in R_i, $i = 1, 2$, and R_1/S_1 is isomorphic to R_2/S_2.*

Proof Choose $s_1 \in S_1$, $r_1 \in R_1$. Since S is a subdirect sum of R_1 and R_2, there exist $s \in S$ and $r_2 \in R_2$ such that $s = r_1 + r_2$. Since $R_1 R_2 = 0$, $s_1 r_1 = s_1(r_1 + r_2) = s_1 s \in S$. Surely, $s_1 r_1 \in R_1$, so $s_1 r_1 \in S \cap R_1 = S_1$. Dually, $r_1 s_1 \in S_1$. Thus S_1 is an ideal in R_1. Similarly, S_2 is an ideal in R_2.

Now define a map $\varphi : R_1/S_1 \to R_2/S_2$ by defining $\varphi(r_1 + S_1) = r_2 + S_2$ where r_2 is chosen so that $r_1 + r_2 \in S$. Suppose, also, that r_2' is such that $r_1 + r'_2 \in S$. Then $(r_1 + r'_2) - (r_1 + r_2) = r'_2 - r_2 \in S_2$ so $r'_2 + S_2 = r_2 + S_2$. Similarly, the choice of r_1 in its coset does not affect the definition of φ, so φ is well defined. From $r_1 + r_2 \in S$, $r'_1 + r'_2 \in S$ follows $(r_1 - r'_1) + (r_2 - r'_2) \in S$, and

$$(r_1 + r_2)(r'_1 + r'_2) = r_1 r'_1 + r_2 r'_2 \in S.$$

Thus φ is a homomorphism. Since φ is clearly onto and one-to-one, φ is an isomorphism from R_1/S_1 to R_2/S_2.

5 The primary decomposition

For any ring R and non-negative integer n, observe that the set of $x \in R$ such that $nx = 0$ forms a two-sided ideal of R. Hence the standard decomposition of an abelian group into a torsion subgroup and a torsion-free factor group, and the decomposition of a torsion abelian group into its primary components, when applied to the additive group of a ring, induce ring decompositions. Because of their frequent application, we now state these observations formally.

1.5.1 THEOREM *If R is a ring and $T = \{x \in R \mid \text{char } x \neq 0\}$, then T is an ideal of R, and R/T contains no non-zero element of non-zero characteristic.*

1.5.2 DEFINITION A ring in which every element has characteristic a power of the prime p is called a *p-ring*.

1.5.3 THEOREM *If R is a ring with a torsion additive group, then R is uniquely expressible as a restricted direct sum of p-rings R_p for different primes p.*

The term "p-ring", together with a proof of the primary decomposition **1.5.3**, can be found in a 1930 paper of K. Shoda [2]. The reader should be cautioned that the notation "p-ring" has also been employed for a certain generalization of Boolean rings, namely for rings R such that $px = 0$ and $x^p = x$ for all $x \in R$. This second use of the term "p-ring" appears to originate with a 1937 paper of N. McCoy and D. Montgomery, who characterized such rings as subdirect sums of copies of $GF(p)$. This class of rings will not be discussed in this book, so this duplication of notation should cause no difficulty for the reader.

Let us observe that the primary decomposition of a torsion ring R induces in a natural way the primary decompositions of subrings and homomorphic images of R. If, for each prime p, we define

$$R_p = \{x \in R \mid \text{char } x = p^n, \text{ some natural number } n\},$$

B

then we obtain the following direct sum decompositions:

$$R \cong \sum_p \oplus R_p, \qquad\qquad \text{and,}$$

if S is a subring of R,

$$S \cong \sum_p \oplus (S \cap R_p), \qquad\qquad \text{or,}$$

if I is an ideal of R,

$$R/I \cong \sum_p \oplus (R_p/(I \cap R_p)).$$

An application of the Chinese Remainder Theorem, moreover, allows us to write an element of R as a sum of natural multiples of itself, where each multiple lies in one of the primary components R_p.

Henceforth, in all results and examples concerning finite rings we shall limit our attention to p-rings. From Theorem **1.5.3** it is clear how to extend these results to general finite rings. The statements of these extensions, however, are often complicated, and provide no additional insight.

6 The circle group

An element x of a ring R, which has an identity 1, is called *invertible* if there is an element $y \in R$ with $xy = yx = 1$. The set of invertible elements of a ring clearly forms a group under multiplication.

Given any ring R one can construct another ring R_1 which has an identity and which contains a subring isomorphic to R. For example, define R_1 to be the set of ordered pairs (x, n) where $x \in R$ and n is an integer. If char $R = m \neq 0$, then n may be chosen to be in the integers mod m. If R is an algebra over a field F, then n may be taken from F. Define $(x_1, n_1) + (x_2, n_2) = (x_1 + x_2, n_1 + n_2)$ and $(x_1, n_1)(x_2, n_2) = (x_1 x_2 + n_1 x_2 + n_2 x_1, n_1 n_2)$. Then R_1 has an identity $(0, 1)$ and a subring $\{(x, 0) \mid x \in R\}$ which is isomorphic to R.

1.6.1   *Let x be a nilpotent element of a ring with identity 1. Then $1 + x$ is invertible, and $(1 + x)^{-1} = 1 + y$ for some $y \in \langle x \rangle$.*

Proof   Let $e = \exp (x)$. Let $y = -x + x^2 - \ldots \pm x^{e-1}$. Then $(1 + y)(1 + x) = (1 + x)(1 + y) = 1$. ($1 + y$ is just the geometric series for $1/1 + x$.)

Now let R be a nil ring, embedded in a ring R_1 with an identity 1. Let $A = \{1 + x \mid x \in R\}$. By **1.6.1** A is closed under inverse, and, since $(1 + x)(1 + y) = 1 + (x + y + xy)$, A is closed under product, and hence is a multiplicative group. From the form of the product, moreover, observe that this group may be defined directly within R by introducing the *circle composition*

$$xoy = x + y + xy.$$

We then have the following result:

1.6.2 THEOREM *Let R be a nil ring. Then R is a group, denoted (R, o), under the circle composition. The identity of (R, o) is the zero of R.*

This result was discovered independently by Perlis [1] and by Mal'cev [1].

It is possible that a ring which is not nil is still a group under the circle composition. An example is the ring $\mathscr{Q}_2$ of rational numbers which as fractions in lowest terms have an even numerator and odd denominator. A ring which forms a group under the circle composition is called a *Jacobson radical* ring. An element of an arbitrary ring which is invertible under the circle composition is called *quasi-regular*.

It may be noted that the circle composition of x and y is sometimes defined to be $x + y - xy$ instead of $x + y + xy$. The next result shows that this operation leads to an isomorphic group.

1.6.3 THEOREM *Let R be a Jacobson radical ring, embedded in a ring R_1 with identity. Let $a \in R_1$ be invertible, with inverse b, and suppose $ax = xa$ for all $x \in R$, and $aR, bR \subseteq R$. Then R is a group $(R, *)$ under the composition*

$$x * y = x + y + axy$$

and $(R, *)$ is isomorphic to (R, o).

Proof For $x \in R$ define $\varphi(x) = ax$. Since a is invertible and aR, $bR \subseteq R$, φ is a one-to-one map of R onto R. Moreover,

$$\varphi(x)o\varphi(y) = ax + ay + a^2xy = \varphi(x * y),$$

so φ is an isomorphism of $(R, *)$ onto (R, o).

Remark The hypothesis $bR \subseteq R$ is required to show that φ is onto. Consider the ring $\mathscr{Q}_2$ embedded in the rationals, with $a = 2$ and $b = 1/2$.

1.6.4 THEOREM *If R is a nilpotent ring, then (R, o) is a nilpotent group.*

Proof The following elementary calculation relating circle and additive commutators shows that an annihilator series for R is a central series for (R, o).

1.6.5 *Suppose* $xo\bar{x} = 0$, $yo\bar{y} = 0$. *Then*

$$\bar{x}o\bar{y}oxoy = (1 + \bar{x})\,(\bar{y}x - x\bar{y})\,(1 + y).$$

Remark The converse of Theorem **1.6.4** is false, as shown by the ring $\mathscr{Q}_2$. In the case of a finitely generated radical ring R, however, nilpotence of the group (R, o) does imply nilpotence of R. See Watters [1] and Kruse [4].

The invertibility of the circle composition in a Jacobson radical ring provides an interesting proof that certain additive subgroups must be ideals. The result is:

1.6.6 THEOREM *Let I be a right (two-sided) ideal of a Jacobson radical ring R. If S^+ is any characteristic subgroup of I^+, then S is a right (two-sided) ideal of R.*

Proof Choose $r \in R$. For $x \in I$ define $\varphi_r(x) = x + xr$. Since $1 + r$ is an invertible element, and I is a right ideal, φ_r is an automorphism of I^+. Hence, for $s \in S^+$, $s + sr \in S^+$, so $sr \in S^+$. Thus S is a right ideal in R. If I is also a left ideal, then, similarly, $rs \in S^+$, so S is a two-sided ideal in R.

Let us now examine some of the relations which hold between the subrings of a Jacobson radical ring R and the subgroups of its circle group. It is easy to check that every subring of R is a subgroup of (R, o), and every two-sided ideal of R is a normal subgroup of (R, o). In general, however, we shall see that not all subgroups of (R, o) need be subrings of R. In fact, a subgroup under o is a subring if and only if it is a subgroup under addition. Moreover, normal subgroups of (R, o), which may or may not be subrings of R, need not be ideals. For example, the center of (R, o) corresponds to the center of R—the set of elements in R which commute with every element of R. This set, which is obviously a subring of R, need not be an ideal. We shall now consider some examples of nilpotent p-rings R which show that one cannot tell from the structure of the circle group alone which subgroups will or will not correspond to subrings.

1 *A fully invariant subgroup which is not a subring* Let R be the ring generated by an element x with $x^2 = 2x$, char $x = 8$. Then (R, o) is abelian of order 8 and type (2, 1). The fully invariant subgroup of (R, o) of elements of orders 1 and 2 consists of 0, $3x$, $4x$, and $7x$. These elements do not form a subring of R.

2   *Elementary abelian groups*   If R is a Jacobson radical ring whose additive and circle groups are elementary abelian p-groups, then all the additive and circle subgroups of a given order in R are indistinguishable up to automorphisms of the groups. We shall, however, find examples of such rings in which the subring structure varies substantially.

Suppose R is a Jacobson radical ring whose additive group is a finite elementary abelian p-group. Then (R, o) is elementary abelian if and only if R is commutative and $x^p = 0$ for all $x \in R$. To prove this consider (R, o) as the multiplicative group of elements $1 + x$ where $x \in R$ and 1 is an identity adjoined to R. Observe that $px = 0$ implies $(1 + x)^p = 1 + x^p$, all $x \in R$.

For the following examples let R be a p-algebra of dimension 3 with a basis $\{a, b, c\}$. Assume that $p \neq 2$.

a   Let R be null. Then every subgroup of R^+ or (R, o) is an ideal of R.

b   Let $a^2 = b$, $a^3 = ac = ca = 0$. Of the $p^2 + p + 1$ subgroups of order p^2, $p + 1$ are subrings (and, by **1.3.7**, ideals). Of the $p^2 + p + 1$ subgroups of order p, $p + 1$ are subrings (and are all ideals).

c   Assume $p > 3$. Let $a^2 = b$, $a^3 = c$, $a^4 = 0$. Then R has only 1 subring (ideal) of order p^2 and $p + 1$ subrings of order p, but only 1 ideal of order p.

d   Let μ be a non-square in $GF(p)$, and let $a^2 = c$, $b^2 = -\mu c$, all other products of basis members 0. Then R has $p + 1$ subrings (ideals) of order p^2, and only 1 subring of order p.

The implications for the structure of a ring of various assumptions concerning its group of quasi-regular elements have been studied by K. E. Eldridge [1–3]. Rings with a cyclic group of quasi-regular elements have been studied by R. W. Gilmer [1], and by Eldridge and I. Fischer [1, 2]. See remark after **4.4.6**.

The characterization of those groups which occur as circle groups of Jacobson radical rings is an interesting problem. For finite groups it suffices, by the primary reduction **1.5.3**, to consider only p-groups. N. H. Eggert [1, 2] has studied the finite abelian groups which occur as circle groups of nilpotent p-*algebras*. If we consider other classes of rings, however, then clearly every abelian group A occurs as the circle group of the null ring with additive group A. L. Kaloujnine [1] has considered the next case:

1.6.7   THEOREM   *Let G be a p-group of class 2, p odd. Then G is isomorphic to the circle group of some nilpotent ring.*

Proof Since $p \neq 2$, the map $x \to x^2$ is one-to-one onto G. Thus the operation "square root" is well defined on G. We now define an addition $+$ and a ring product $\cdot$ on G by

$$x \cdot y = xy\, x^{-1}y^{-1}$$
$$x + y = xy\, \sqrt{y \cdot x}.$$

We must verify that under $+$ and $\cdot$ G is a ring. First note that for all $x, y, z \in G$, $\sqrt{y \cdot x}$ is in the commutator subgroup of G, and hence is central in G, and that $(x \cdot y) \cdot z = x \cdot (y \cdot z) = e$, where e is the identity of G.

(1) Associativity of $+$:

$$\begin{aligned}
(x + y) + z &= xy\, \sqrt{y \cdot x}\, \sqrt{z \cdot (xy\, \sqrt{y \cdot x})} \\
&= xyz\, \sqrt{y \cdot x}\, \sqrt{(z \cdot xy)} \text{ since } \sqrt{y \cdot x} \text{ is central} \\
&= xyz\, \sqrt{y \cdot x}\, \sqrt{z \cdot x}\, \sqrt{z \cdot y}.
\end{aligned}$$

Also $x + (y + z) = xyz\, \sqrt{z \cdot y}\, \sqrt{yz \cdot x} = xyz\, \sqrt{y \cdot x}\, \sqrt{z \cdot x}\, \sqrt{z \cdot y}.$

(2) Let e be the identity of G. e is clearly a zero under $+$.

(3) The inverse x^{-1} of an element $x \in G$ satisfies $x + x^{-1} = e$.

(4) Commutativity of $+$:

$$\begin{aligned}
x + y &= xy\, \sqrt{y \cdot x} = xy\, \sqrt{yxy^{-1}x^{-1}} \\
&= xy\, \sqrt{yxy^{-1}x^{-1}}\, \left(\sqrt{xyx^{-1}y^{-1}}\, \sqrt{xyx^{-1}y^{-1}}\, yxy^{-1}x^{-1} \right) \\
&= yx\, \sqrt{xyx^{-1}y^{-1}} = y + x.
\end{aligned}$$

(5) Left distributivity:

$$\begin{aligned}
x \cdot (y + z) &= x \cdot (yz\, \sqrt{z \cdot y}) = x \cdot (yz) = (x \cdot y)\, (x \cdot z) \\
&= (x \cdot y)\, (x \cdot z)\, \sqrt{(x \cdot z) \cdot (x \cdot y)} \\
&= (x \cdot y) + (x \cdot z).
\end{aligned}$$

Right distributivity is dual.

Thus G is a nilpotent ring of exponent 3 under $+$ and $\cdot$. Let $x * y = x + y + \frac{1}{2}(x \cdot y)$. By **1.6.3** it is sufficient to show that $(G, *)$ is isomorphic to the original group G. But $\frac{1}{2}(x \cdot y) = \sqrt{x \cdot y}$, so

$$x * y = xy\, \sqrt{x \cdot y}\, \sqrt{y \cdot x} = xy.$$

This completes the proof of **1.6.7**.

We shall conclude this section by obtaining the following restriction on those p-groups which can occur as circle groups.

1.6.8 THEOREM *Let G be a p-group. If G occurs as the circle group of a nilpotent ring, then G has a central series*

$$G = Z_0 \supset Z_1 \supset \ldots \supset Z_c = 1$$

in which $[Z_{i-1}:Z_i] \geq p^2$ for $1 \leq i < c$.

Remark 1 Note that the theorem makes no assertion concerning $[Z_{c-1}:Z_c] = |Z_{c-1}|$. One can trivially see that $|Z_{c-1}| = p$ can occur by considering the null ring of order p.

Remark 2 To show that the bound on indices in **1.6.8** cannot be improved, consider the following example. Let R be the ring with a basis $\{a, b\}$ such that char $a = $ char $b = p^n$, and $a^2 = ab = pa$, $ba = b^2 = pb$. The terms of the upper central series of the circle group of R coincide with the multiples of R of the form $p^k R$, $1 \leq k \leq n$, so all terms of the upper central series ascend by steps of p^2.

Remark 3 Combining Theorem **1.6.8** with Theorem **1.6.7** for $p \neq 2$, or with a direct construction for $p = 2$, we immediately obtain:

1.6.9 COROLLARY *All groups of orders p, p^2, and p^3 occur as circle groups of nilpotent rings. A group of order p^4 occurs as a circle group if and only if it is abelian or has class 2.*

The theorem will be obtained as a corollary to the following lemma.

1.6.10 LEMMA *Let R be a nilpotent p-ring with* exp $R = n$ *and* $[R^{n-2}:R^{n-1}] = p$, $n \geq 3$. *Then R^{n-2} is in the center of R.*

Proof Let $b = r_1 r_2 \ldots r_{n-2}$ be a product of length $n - 2$ in R, with $b \notin R^{n-1}$. Since $[R^{n-2}:R^{n-1}] = p$, it follows that $pb \in R^{n-1}$, and b together with R^{n-1} spans R^{n-2}. Since R^{n-1} annihilates R, R^{n-1} is in the center of R. Assume R^{n-2} is not in the center. Thus there is an element $a \in R$ with $ab \neq ba$. Hence either $ab \neq 0$ or $ba \neq 0$. To be definite suppose $ab \neq 0$.

We shall now show by a finite induction that $b \equiv xa^{n-2} \pmod{R^{n-1}}$ for some integer $x \not\equiv 0 \pmod{p}$. Suppose, for some integer k, $0 \leq k \leq n - 3$, that

(1) $\qquad b \equiv x_k a^k r_1 r_2 \ldots r_{n-k-2} \pmod{R^{n-1}}$,
$$\text{some integer } x_k \not\equiv 0 \pmod{p}.$$

Since R^{n-1} annihilates R, (1) implies $ab = (x_k a^{k+1} r_1 \ldots r_{n-k-3}) r_{n-k-2}$. Since $0 \neq ab \in R^{n-1}$, it follows that $s = x_k a^{k+1} r_1 \ldots r_{n-k-3} \notin R^{n-1}$. But $s \in R^{n-2}$, so $b \equiv ys \pmod{R^{n-1}}$ for some integer $y \not\equiv 0 \pmod{p}$. Setting $x_{k+1} = yx_k$ we obtain (1) with k replaced by $k + 1$.

Condition (1) trivially holds for $k = 0$, so by induction we obtain $b \equiv x_{n-2} a^{n-2} \pmod{R^{n-1}}$ for some integer $x_{n-2} \not\equiv 0 \pmod{p}$. Since

R^{n-1} annihilates R we then have $ab = x_{n-2}a^{n-1} = ba$, which contradicts the assumption that $ab \neq ba$. Hence R^{n-2} is central in R.

Proof of **1.6.8**   Let R be a nilpotent p-ring with circle group G.
Since R is nilpotent, the powers of R form a finite descending central
series for G. We shall select a subset

$$R = Z_0 \supset Z_1 \supset \ldots \supset Z_c = 0$$

which forms a central series for G and for which $[Z_{i-1}:Z_i] \geq p^2$,
$1 \leq i < c$. To do so let us define $Z_0 = R$ and proceed by induction.
Suppose $Z_i = R^k \neq 0$. If $[R^k:R^{k+1}] \geq p^2$, or if $R^{k+1} = 0$, define
$Z_{i+1} = R^{k+1}$. If $[R^k:R^{k+1}] = p$ and $R^{k+1} \neq 0$, then application of the
lemma to the ring R/R^{k+2} shows that R^k/R^{k+2} is central in R/R^{k+2}.
In this case define $Z_{i+1} = R^{k+2}$. This process terminates when $Z_i = 0$.
and the description of the desired central series is then complete.

CHAPTER II

Examples of nilpotent rings

1 Representations and rings of matrices

LET M BE AN abelian group of finite rank, with a basis $\{a_1, \ldots, a_n\}$. If R is any ring, associative or not, such that $R^+ = M$, then R is determined up to isomorphism by specifying the products of basis elements as integer linear combinations of basis elements

$$a_i a_k = \sum_j r_{ijk} a_j, \tag{1}$$

$1 \leq i, j, k \leq n$. To consider condition (1) geometrically let us introduce a Kronecker product $T = M \otimes M \otimes M$. T can be defined as an abelian group of rank n^3, with a basis $t_{ijk} = a_i \otimes a_j \otimes a_k$, $1 \leq i, j, k \leq n$, such that the characteristic of t_{ijk} is the greatest common divisor of the characteristics of a_i, a_j, and a_k. The ring R corresponds to one point r in T, given by $r = \sum_{i,j,k} r_{ijk} t_{ijk}$.

Next consider a second ring R' such that $(R')^+ = M$. R' corresponds to some point $r' \in T$. Further R' is isomorphic to R if and only if there is a change of basis of M which takes each r_{ijk} onto r'_{ijk}. This change of basis may be expressed as an invertible integer matrix U written on the left of $x \in M$, and with column i of U interpreted modulo the characteristic of a_i. A direct calculation then gives

$$r' = (U \otimes (U^{-1})^t \otimes U)\, r$$

where $(U^{-1})^t$ denotes the transpose of U^{-1}, and the Kronecker product of matrices is defined in the usual way.

The relation between rings over an additive group M and the sets of structure constants r_{ijk} of (1) has been studied by several authors. It appears, for example, in a 1948 paper of R. A. Beaumont [1]. The isomorphism problem formulated in terms of Kronecker products appears in B. R. Toskey [1], and in a slightly more general form in K. Baumgartner [1].

From now on we consider only associative rings. From condition (1) let us introduce matrices $X^{(k)}$, $1 \leq k \leq n$, by defining $X^{(k)}{}_{ij} = r_{ijk}$. These matrices span a subring of the ring of integer $n \times n$ matrices for which column i is interpreted modulo the characteristic of a_i. This subring, moreover, is a homomorphic representation of the ring R, and is faithful if and only if $\mathfrak{A}_r(R) = 0$. In case $\mathfrak{A}_r(R) \neq 0$ we may obtain a faithful representation of R by first adjoining an identity to R and then performing the above construction. In the case when R is finite, moreover, it is sufficient by **1.5.3** to suppose that R is a p-ring. In this case we have the following result of T. Szele [1]:

2.1.1 THEOREM *Let R be a ring of type $_p(m_1, \ldots, m_n)$. Then R is isomorphic to a ring of integral $(n + 1) \times (n + 1)$ matrices whose i^{th} columns are interpreted mod p^{m_i}, $1 \leq i \leq n + 1$, with $m_{n+1} = \max\limits_{1 \leq i \leq n} m_i$.*

Let us note that the construction of matrices $X^{(k)}$ from the structure constants r_{ijk} of (1) does not depend on the associativity of R. In fact, R is associative if and only if the additive group of matrices spanned by the $X^{(k)}$ is closed under matrix multiplication. In machine computation this closure criterion is a convenient, systematic procedure for checking the associative law.

In the case of a finite-dimensional algebra A over a field F the representation analogous to **2.1.1** of course has its entries in the field F. If A is nilpotent of exponent e, then we may choose a basis for A by first choosing a basis for A^{e-1}, extending this to a basis for A^{e-2}, and so on. We thus obtain the 1903 result of J. B. Shaw [1], [2]:

2.1.2 THEOREM *Let A be an algebra of dimension n over a field F. Then A is isomorphic to a subalgebra of the $(n + 1) \times (n + 1)$ matrices over F. If A is nilpotent, the matrices may be chosen to be strictly lower triangular. The entries in the last row of each matrix give the description of the corresponding element of A as a linear combination of the given basis members. The last column of each matrix is 0.*

Because of Theorem **2.1.2** there is some interest in the study of the algebra of all strictly lower triangular $n \times n$ matrices over a field F. Such an algebra is called the *total nilpotent algebra of degree n over F.* This nomenclature and the observations we shall give are due to R. Dubisch and S. Perlis [1].

Let T be a total nilpotent algebra of degree n, denoted symbolically as the set of matrices

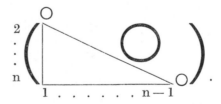

The powers of T are then, for $1 \leq k < n$,

$$T^k = \left\{ \begin{matrix} k+1 \\ \vdots \\ n \end{matrix} \right\}$$

Thus T has exponent n. Further, the upper and lower annihilator series for T coincide. We next define the *left annihilator series* $L_1, L_2, \ldots$ for a ring R as follows. $L_1 = \mathfrak{A}_l(R)$. Then L_1 is a two-sided ideal of R. Suppose L_k is defined. Then L_{k+1} is defined by $L_{k+1} \supseteq L_k$ and $L_{k+1}/L_k = \mathfrak{A}_l(R/L_k)$. The right annihilator series $R_1, \ldots$ is defined dually. Then, for $R = T$,

A complete description of the ideal structure of T may be found in Dubisch and Perlis [1].

2 Free rings

In the last section we found that any ring of finite rank is isomorphic to a subring of a matrix ring. In this section, dually, we introduce a class of rings called free rings such that any ring is a homomorphic image of some free ring.

Let S be a set. The elements of S will be called *indeterminates*. An ordered n-tuple of indeterminates is called a *monomial*. The number n of indeterminates in a monomial is its *length*. The *product* of two monomials $a_1 a_2 \ldots a_n$ and $b_1 b_2 \ldots b_m$, is the monomial $a_1 a_2 \ldots a_n b_1 b_2 \ldots b_m$. The *free ring* $F[S]$ on the set S is the set of all linear combinations of monomials on S with integer coefficients, and with the product of elements obtained by linearity from the cocatenation of monomials.

2.2.1 THEOREM *Let X be a subset of a ring R such that $R = \langle X \rangle$. If S is a set of the same cardinality as X, then R is a homomorphic image of the free ring $F[S]$ under the homomorphism determined by $s \to x$, all $s \in S$, where s and x correspond under a one-to-one correspondence of S and X.*

Proof See, for example, Redei [4], §49, or Kurosh [2], §23.

2.2.2 DEFINITION   The *free nilpotent ring* of exponent e on a set S is $FN_e[S] = F[S]/I$ where I is the ideal in $F[S]$ generated by all monomials of length at least e.

Now suppose R is a nilpotent ring with exp $R = e$, and φ is a natural homomorphism of a free ring F onto R. Then if $w \in F$ is any monomial of length e or more, then $\varphi(w) = 0$. Thus we have

2.2.3 THEOREM   *Let R be a nilpotent ring with exp $R = e$; let X be a subset of R with $R = \langle X \rangle$, and let S be a set with the same cardinality as X. Then there is a homomorphism of $FN_e[S]$ onto R with $x \to s$ for all corresponding $x \in X$, $s \in S$.*

Finally, let us note that if char $R = m \neq 0$, then, in a way similar to the above, we may define a free ring of characteristic m which maps onto R. If R is an algebra over a field F, then we may define a free algebra over F which maps onto R.

By counting the number of monomials of each length in A, a free nilpotent algebra of exponent e on a finite set S, where $|S| = s > 1$, we obtain

$$\dim A = s + s^2 + \ldots + s^{e-1} = \frac{s^e - 1}{s - 1} - 1.$$

Since, moreover, $s = \dim (A/A^2)$, it follows that a nilpotent algebra B of exponent e is free if and only if $\dim B = \dfrac{d^e - 1}{d - 1} - 1$, where $d = \dim (B/B^2)$. These observations are the essential context of two papers of G. Scorza [3, 8].

It may be of interest to note that, whereas a subgroup [subring] of a free group [free ring] must be free, a subring of a free nilpotent ring need not be free. For example, if R is a free nilpotent ring with exp $R \geq 5$ then R^2 is not free.

3 The construction of rings by generators and relations

This section presents a determination of some classes of nilpotent p-rings of small additive rank. These include the nilpotent rings with cyclic additive group, selected rings generated by a single element, and the nilpotent algebras of dimension 3 over a field.

The general procedure is to select a basis with specified properties **for** the additive group of the ring (or algebra) R under consideration.

The members of the basis are generally denoted a, b, c, ... The ring (algebra) R is then determined by specifying the products of the basis elements as integer (field) linear combinations of the basis elements. Products of basis elements which are not explicitly specified are to be assumed zero. The rational integers (field elements) appearing in the products of basis elements are known as the *structure constants* of R. Greek letters will denote structure constants. The choice of structure constants in R is of course limited by the specified properties of R, and by the ring axioms. For example, we frequently use the fact that the characteristic of a product must divide the characteristics of both factors.

To determine which sets of structure constants define isomorphic rings we consider changes of basis for R. The image of $x \in R$ we generally denote x'. Finally, to show that our cases are mutually non-isomorphic we must consider a general change of basis which preserves the specified properties of R and its basis. In doing so we frequently ignore changes of basis which do not affect the structure constants. For example, if a is a basis member for R, $a \notin R^2$, and $x \in \mathfrak{A}(R)$ such that char x divides char a, then the change of basis $a \to a' = a + x$, all other basis elements fixed, does not change the structure constants and is always ignored.

Throughout this section we shall make frequent use of the following application of **1.3.8**. The idea used in its proof will be exploited more fully in Chapter IV to prove a general basis theorem.

2.3.1 LEMMA   *Let R be a nilpotent ring. Suppose R/R^2 is cyclic. Then there exists $a \in R$ such that $R = \langle a \rangle$.*

Proof   Suppose $a + R^2$ is a generator for $(R/R^2)^+$. Then $\langle a \rangle + R^2 = R$, so by **1.3.8**, $\langle a \rangle = R$.

The nilpotent rings generated by a single element are called *power rings*, since a generator together with its powers spans the additive group of the ring. The analogous notation *power algebra* appears well established in the older literature, appearing, for example, as early as a 1921 book of Scorza [1].

The simplest examples of power rings are the rings with finite cyclic additive group, which we now characterize.

2.3.2 THEOREM   *There are exactly $n + 1$ mutually non-isomorphic cyclic rings of order p^n, $n \geq 1$. Each contains an element a, char $a = p^n$, such that $a^2 = p^i a$ for some integer i, $0 \leq i \leq n$.*

Remark   The ring with $i = 0$ is clearly isomorphic to the ring of integers mod p^n, and is not nilpotent, while the rings with $i > 0$ are all nilpotent.

Proof Let R be a cyclic ring of order p^n, with a generator b for R^+. Let $b^2 = \alpha p^i b$, where α is an integer, $\alpha \not\equiv 0 (\mathrm{mod}\ p)$, and $0 \leq i \leq n$. Let $a = \beta b$, where β is an integer with $\alpha \beta \equiv 1(\mathrm{mod}\ p^n)$. Then a also generates R^+, and $a^2 = \beta^2 \alpha p^i b = \beta p^i b = p^i a$. Since, finally, char R^2 is an isomorphism invariant, different values of i define non-isomorphic rings.

Historical remark The classification of finite cyclic rings appears many times in the literature. Some authors fail to make use of the primary decomposition, with a resulting substantial increase in complexity of the statements of their results and proofs.

2.3.3 THEOREM *Let R be a nilpotent ring of order p^2 which is not cyclic. Then there is a basis $\{a, b\}$ for R such that* char $a =$ char $b = p$ *and one of the following holds.*
(1) $a^2 = b^2 = ab = ba = 0$.
(2) $a^2 = b$, $a^3 = 0$.

Proof Since R is not cyclic, R has type $_p(1, 1)$. If $R^2 = 0$, then (1) holds. If $R^2 \neq 0$, then $|\ R/R^2\ | = p$, so, by **2.3.1** there exists $a \in R$ such that $\langle a \rangle = R$. Hence a^2 is an additive generator of R^2, and the nilpotence of a implies that $a^3 = 0$. Hence (2) holds.

A group generated by a single element is completely determined by its order. Power rings, however, are not completely determined by order, or even by additive group type. We now determine some of the simplest power rings. The first result, which follows directly from **1.2.4**, is due to Frobenius [2]:

2.3.4 THEOREM *Let A be a nilpotent algebra of dimension n over a field F, generated by a single element a. Then $\{a, a^2, \ldots, a^n\}$ forms a basis for A, and $a^{n+1} = 0$. Thus, up to isomorphism, there is only one power algebra of dimension n over F.*

2.3.5 THEOREM *Let R be a power ring of type $_p(2,1,1)$. Then R contains a basis $\{a, b, c\}$ with $p^2 a = pb = pc = 0$, $a^2 = b$, $a^3 = c$, $a^4 = \alpha pa$, where α satisfies one of the following conditions. These cases and their subcases are mutually non-isomorphic.*
(1) $p \not\equiv 1 \ (\mathrm{mod}\ 3)$. $\alpha = 0$ or $\alpha = 1$.
(2) $p \equiv 1 \ (\mathrm{mod}\ 3)$. $\alpha = 0, 1, \mu, \nu$, where $1, \mu, \nu$ are representatives of the cosets of $(GF^*(p))^3$ in $GF^*(p)$.

Proof Let $R = \langle a \rangle$. Since char $R =$ char a, char $a = p^2$. Let $a^2 = b$, $a^3 = c$. Since $R/\langle pa \rangle$ is a power p-algebra of dimension 3, the images of a, b, c span $(R/\langle pa \rangle)^+$. Moreover, $pR^2 = 0$ since, by **1.3.5**, $R^2 \subseteq \{x \in R \mid px = 0\}$. It follows that $\{a, b, c\}$ forms a basis for R^+, and that $a^4 \in \langle pa \rangle$. Let $a^4 = \alpha pa$. Consider the change of basis $a' = \beta a$, $b' = a'^2$, $c' = a'^3$ where $\beta \not\equiv 0 \pmod{p}$. Then $a^4 = \beta^4 a^4 = \beta^3 \alpha pa'$. If $p \not\equiv 1 \pmod 3$, then β^3 runs over the integers mod p as β does, and so a change of basis gives $\alpha = 0$ or $\alpha = 1$. This is case (1). If $p \equiv 1 \pmod 3$, then either $\alpha = 0$ or $\beta^3\alpha$ may be made one of the fixed representatives 1, μ, ν of the cosets of $(GF^*(p))^3$ in $GF^*(p)$. Thus (2) holds.

Additional classes of power rings will be characterized in Chapter V. We conclude this section with the following

2.3.6 THEOREM *Let A be a nilpotent algebra of dimension 3 over a field F. Then A has a basis $\{a, b, c\}$, with $c \in \mathfrak{A}(A)$, such that one of the following conditions holds. All of the algebras described in these cases are mutually non-isomorphic.*
(1) $a^2 = ab = ba = b^2 = 0$ *(null algebra)*.
(2) $a^2 = b^2 = 0$, $ab = -ba = c$.
(3) $a^2 = c$, $ab = ba = b^2 = 0$ *(directly reducible)*.
(4) $a^2 = c$, $ab = ba = 0$, $b^2 = \gamma c$, *where γ is a predetermined representative of its coset of $(F^*)^2$ in F^*.*
(5) $a^2 = ab = c$, $ba = 0$, $b^2 = \varphi c$, *some $\varphi \in F$.*
(6) $a^2 = b$, $a^3 = c$, $a^4 = 0$ *(power algebra)*.

Proof If dim $A^2 = 2$ then, by **2.3.1** and **2.3.4**, (6) holds. If $A^2 = 0$ then (1) holds. Hence we shall assume dim $A^2 = 1$, which implies $A^3 = 0$.

First suppose $x^2 = 0$ for all $x \in A$. Choose a, b linearly independent mod A^2. From $a^2 = b^2 = (a + b)^2 = 0$ follows $ab = -ba$. Since $A^2 \neq 0$, $ab = c \neq 0$. Hence (2) holds.

Now suppose $a \in A$ has $a^2 = c \neq 0$. Choose $b \notin \langle a \rangle$. Then $\{a, b, c\}$ is a basis for A. Let $ba = \delta c$. Then $(b - \delta a) a = 0$, so we can choose b to make $ba = 0$. Let $ab = \alpha c$, $b^2 = \varphi c$. Suppose $\alpha = 0$. If $\varphi = 0$, then (3) holds. If $\varphi \neq 0$, and if we replace b by $b' = \psi b$, $0 \neq \psi \in \mathfrak{F}$, then φ becomes $\varphi' = \psi^2 \varphi$. Thus we may force φ to be any predetermined coset representative of $(F^*)^2$ in F^*. Thus (4) holds. If, on the other hand, $\alpha \neq 0$, then replacing b by $b' = \alpha^{-1}b$ we obtain case (5).

The cases described in (1)—(6) are clearly mutually non-isomorphic, except that the structure constant φ of case (5) might change within an isomorphism class. To show that φ cannot be changed, consider a change of basis of A

$$a' = \zeta a + \eta b, \quad b' = \kappa a + \lambda b,$$

with $c' = a'^2 = a'b' \neq 0$, $b'a' = 0$, and $\zeta\lambda - \eta\kappa \neq 0$. These conditions imply

$$\zeta^2 + \zeta\eta + \eta^2\varphi = \zeta\kappa + \zeta\lambda + \eta\lambda\varphi \neq 0 \tag{1}$$

and
$$\zeta\kappa + \eta\kappa + \eta\lambda\varphi = 0. \tag{2}$$

(1) implies $\zeta(\zeta^2 + \zeta\eta + \eta^2\varphi) = \zeta^2\kappa + \zeta(\zeta\lambda) + \eta\varphi(\zeta\lambda) \tag{3}$

and (1)—(2) gives $\zeta\lambda = \zeta^2 + \zeta\eta + \eta^2\varphi + \eta\kappa \tag{4}$

Substituting (4)

   into (3) gives $0 = (\kappa + \eta\varphi)(\zeta^2 + \zeta\eta + \eta^2\varphi).$

(1) then implies $\kappa + \eta\varphi = 0$, and (2) becomes $\eta\varphi(\lambda - \zeta - \eta) = 0$. Thus one of $\kappa = \eta = 0$, $\kappa = \varphi = 0$, or $\lambda = \zeta + \eta$ holds. In none of these cases can the structure constant φ be changed.

4 Modular group algebras of p-groups

If G is a group and F a field, the *group algebra* $F[G]$ *of* G *over* F is defined to be the algebra over F with a basis consisting of all $g \in G$, and multiplication extending linearly the multiplication of G. If G is a finite group, then $F[G]$ is a semi-simple ring if and only if the characteristic of F does not divide the order of G (Maschke [1]). To obtain an example of a nilpotent ring, consider the case when G is a finite p-group and char $F = p$. The following result is due to L. E. Dickson [2]:

2.4.1 THEOREM *Let* G *be a group of order* p^n *and* F *a field of characteristic* p. *The radical* $N[G]$ *of* $F[G]$ *is of dimension* $p^n - 1$ *over* F *and has as a basis all elements of the form* $g - 1$ *where* $g \in G$, $g \neq 1$. *An element* $x = \sum \alpha_i g_i$, *where* $\alpha_i \in F$, $g_i \in G$, *lies in* $N[G]$ *if and only if* $\sum \alpha_i = 0$. *The semi-simple algebra* $F[G]/N[G]$ *is isomorphic to* F.

Proof The elements $g - 1$ for $g \in G$, $g \neq 1$, clearly are linearly independent, and span an ideal N of $F[G]$ of dimension $p^n - 1$. Moreover, an element $x = \sum \alpha_i g_i \in F[G]$ lies in N if and only if $\sum \alpha_i = 0$, and clearly the algebra $F[G]/N$ is not nilpotent and has dimension 1, so is isomorphic to F. Thus to complete the proof we need only establish that the ideal N is nilpotent. We shall give two proofs of the nilpotence of N, one in the context of finite groups, and the other in the context of finite-dimensional algebras.

Proof 1 The right regular representation of $F[G]$ represents the group G as a group of linear transformations over the field F of characteristic p. If the natural basis $\{g \mid g \in G\}$ is chosen for $F[G]$ then the group elements g in this representation correspond to matrices of 0's and 1's, and so are matrices over the prime field $GF(p)$. It is easily verified that, for all $m \geq 2$, the set S of lower triangular $m \times m$ matrices over $GF(p)$ with 1's on the diagonal forms a Sylow p-subgroup of the general linear group $GL(m, p)$ over $GF(p)$. Since all Sylow p-subgroups of a finite group are conjugate and G is contained in some Sylow p-subgroup of $GL(p^n, p)$ there exists a basis for $F[G]$ such that the elements of G correspond to matrices in S. In this basis the elements of N correspond to strictly lower triangular matrices, which form a nilpotent subalgebra, so N is nilpotent.

Proof 2   Since char $F = p$, the binomial theorem gives $(g - 1)^{p^n} - g^{p^n} - 1 - 0$ for all $g \in G$. Thus N has a basis of nilpotent elements. That N is nilpotent follows from a classical result of Wedderburn [4], which we now prove:

2.4.2 THEOREM   *Let A be an algebra over a field F with a finite basis of nilpotent elements. Then A is nilpotent.*

Proof   Let F' be an algebraically closed field containing F, A' the Kronecker product over F of A and F'. Regard A as a subring of A'. A' has a nilpotent basis. Let N' be the radical of A'. If A' is not nilpotent, then the semi-simple algebra $A'/N' \neq 0$ has a nilpotent basis. Since F' is algebraically closed and A' is finite dimensional over F', A'/N' is isomorphic to a direct sum of complete matrix rings over F'. With this isomorphism, introduce an additive linear function, trace, mapping $A'/N' \to F'$. Nilpotent matrices over F' have trace 0; so A'/N' has a basis consisting of elements of trace 0; so trace is identically 0 on A'/N'. But the matrix unit e_{11} has trace 1, a contradiction. So $A' = N'$; A' is nilpotent, hence A is nilpotent.

Remark   If G is a group of order 4 and F is a field of characteristic 2 then the radical N of $F[G]$ has dimension 3 and so is isomorphic to one of the algebras described in **2.3.6**. If G is cyclic then N is a power algebra, and satisfies (6) of **2.3.6**. If G is the Klein 4-group then N satisfies (2) of **2.3.6**.

An interesting open problem is whether a finite p-group G is determined up to isomorphism by the algebra structure of $F[G]$, where F is a field of characteristic p. This has been answered in the affirmative for groups of order 8 by D. G. Coleman [1] and, later, by R. Holvoet [1], for groups of order p^4 by D. S. Passman [1], and for finite abelian p-groups by W. E. Deskins [1]. We conclude this section with a proof of Deskins' result. The proof is from Coleman [1].

C

2.4.3 THEOREM *Let G and H be finite abelian p-groups, F a field of characteristic p. If $F[G] \cong F[H]$ then $G \cong H$.*

Proof For any set X of group algebra elements, let $X^p = \{x^p \mid x \in X\}$. We shall prove the theorem by induction on $|G| = |H|$. For $|G| = |H| = 1$ the result is trivial.

Clearly any isomorphism of $F[G]$ onto $F[H]$ induces an isomorphism of $(F[G])^p$ onto $(F[H])^p$. But $(F[G])^p = F^p[G^p]$ and $(F[H])^p = F^p[H^p]$, where F^p is the field of pth powers of members of F. Hence the induction hypothesis implies $G^p \cong H^p$, since $|G^p| = |H^p| < |G|$. For finite abelian p-groups, however, the conditions $G^p \cong H^p$ and $|G| = |H|$ are enough to ensure that $G \cong H$.

CHAPTER III

Families and capability

1 The family classification

AN IMPORTANT GOAL in the classification of mathematical systems is to find useful properties which reflect the structure of the systems under consideration. Ideally, enough properties may be found so that the systems are uniquely determined by these properties. But even a partial set of properties will often shed light on the structure of the system, or aid in solution of a problem by restricting the possible configurations which may appear.

The structure of the additive group is certainly an important property for the classification of rings. While the additive group structure indicates many things about the ring, it usually gives little information about the second operation in a ring, multiplication. This chapter introduces a classification of rings which closely reflects the multiplicative structure of the ring. Observe that multiplication in a ring R can be regarded as a function whose domain is pairs of elements in the ring $R/\mathfrak{A}(R)$ and whose range is the ring R^2. In order to classify rings according to the similarity of the multiplication functions they induce, we make the following

3.1.1 DEFINITION Rings R and S are in the same *family*, denoted $R \overset{F}{\leftrightarrow} S$, if
(1) There is an isomorphism $\varphi\colon R/\mathfrak{A}(R) \to S/\mathfrak{A}(S)$, and
(2) There is an isomorphism $\psi\colon R^2 \to S^2$, such that
(3) If $r_i + \mathfrak{A}(R) \overset{\phi}{\to} s_i + \mathfrak{A}(S)$, $i = 1, 2$, then $r_1 r_2 \overset{\psi}{\to} s_1 s_2$.

From the definition of family follow immediately:

3.1.2 $\overset{F}{\leftrightarrow}$ *is an equivalence relation.*

3.1.3 *The set of all null rings constitutes a single family.*

3.1.4 *If R is a ring and N a null ring, then $R \oplus N \overset{F}{\leftrightarrow} R$.*

3.1.5 *If ring R is nilpotent, then $R \overset{F}{\leftrightarrow} S$ implies ring S is nilpotent.*

Observe that, for the class of rings R with $\mathfrak{A}(R) = 0$ and $R^2 = R$, in particular, for all rings with an identity, the family relation is equivalent to isomorphism. Although stated generally, the results of this chapter are of interest mainly for nilpotent rings.

Let us recall the analogy between the theory of nilpotent rings and the theory of nilpotent groups, which was introduced in §1.3. In this analogy an abelian group corresponds to a null ring, the center of a group to the anninilator of a ring, and the derived group to the square of a ring. Under this analogy the definition of families of rings and several of the early results correspond to the theory of families of groups presented by Philip Hall [3]. In the development of the ring theory, however, it soon becomes apparent that the additional structure available in a ring allows a more extensive development of the ideas than is possible for groups.

In §2 of this chapter we investigate the relations which hold among the rings of a single family. Our main interest is to find relations between finite nilpotent p-rings and p-algebras. Given a p-algebra, we describe a procedure for constructing all p-rings lying in the same (ring) family with the algebra, which uses only the structure of the algebra and its automorphism group. This construction is of substantial value in determining the nilpotent rings of small orders.

In §3 this procedure is illustrated by the classification of the nilpotent rings of order p^3. In §4 another use of the family relation is provided by the study of nilpotent rings with chain conditions.

§5 considers the problem of constructing all families of rings R for which $R/\mathfrak{A}(R) \cong S$ for a fixed ring S. The method is the study of a certain quotient ring (the Schur Multiplier) in a free ring. Not every ring S can occur as $R/\mathfrak{A}(R)$ for some ring R. A ring S which can so occur is called *capable*. We shall obtain a necessary and sufficient condition that a ring be capable. The methods and results of §5 are analogous to the group theory of Philip Hall [3].

§6 determines several explicit conditions for the capability of a ring in terms of its subrings. Here stronger relations are possible than occur for groups. It is shown, for example, that capability of a direct sum of finite rings implies capability of each summand. The analogous group statement fails to hold. The chapter concludes by determining the capability of some special classes of nilpotent rings. These results show that the structure of a nilpotent ring modulo its annihilator is severely limited. There are, for example, $3p + 15$ nilpotent, not directly reducible rings of order p^3, p an odd prime. Of these rings, only 6 are capable. Hence the study of capable rings should lead to significant insights into the properties of nilpotent rings.

2. On the rings in a family.

3.2.1 THEOREM   *Let R_1 be a subring of a ring R. If $R = R_1 + \mathfrak{A}(R)$, then $R \overset{F}{\leftrightarrow} R_1$. The converse holds for finite rings.*

The proof requires a lemma.

3.2.2 LEMMA   *If $R = R_1 + \mathfrak{A}(R)$, then $\mathfrak{A}(R_1) = R_1 \cap \mathfrak{A}(R)$.*

Proof $\mathfrak{A}(R_1) \supseteq R_1 \cap \mathfrak{A}(R)$ is clear. Choose $x \in \mathfrak{A}(R_1)$. To prove $x \in \mathfrak{A}(R)$ choose $y \subset R$ and write $y = r_1 + a$, where $r_1 \in R_1$, $a \in \mathfrak{A}(R)$. Then $xy = xr_1 + xa = 0 + 0$. Similarly $yx = 0$, so $x \in \mathfrak{A}(R)$.

Proof of **3.2.1**   Suppose $R = R_1 + \mathfrak{A}(R)$. Define a map φ: $R_1/\mathfrak{A}(R_1) \to R/\mathfrak{A}(R)$ by $\varphi(x + \mathfrak{A}(R_1)) = x + \mathfrak{A}(R)$, all $x \in R_1$. By **3.2.2** φ is well defined and one-to-one. φ is clearly a homomorphism. Since $R = R_1 + \mathfrak{A}(R)$, φ is onto. Thus φ is an isomorphism. From $R = R_1 + \mathfrak{A}(R)$ follows $R^2 = R_1^2$. Let $\psi: R_1^2 \to R^2$ be the identity isomorphism. Clearly φ and ψ satisfy condition (3) of **3.1.1**, so $R_1 \overset{F}{\leftrightarrow} R$.

For the converse suppose R_1 is a subring of a finite ring R, with $R \overset{F}{\leftrightarrow} R_1$. Since $\mathfrak{A}(R_1) \supseteq R_1 \cap \mathfrak{A}(R)$, $R/\mathfrak{A}(R) \cong R_1/\mathfrak{A}(R_1)$ is a homomorphic image of $R_1/(R_1 \cap \mathfrak{A}(R)) \cong (R_1 + \mathfrak{A}(R))/\mathfrak{A}(R)$. Since R is finite, $R = R_1 + \mathfrak{A}(R)$ follows.

3.2.3 COROLLARY   *Let I be an ideal of a ring R. If $I \cap R^2 = 0$, then $R \overset{F}{\leftrightarrow} R/I$. The converse holds for finite rings.*

Proof   Suppose $I \cap R^2 = 0$. By **1.4.1**, the mapping $x \to (x + I, x + R^2)$ expresses R as a subdirect sum of R/I and the null ring $R/R^2 = N$. By **3.1.4** $R/I \overset{F}{\leftrightarrow} R/I \oplus N$, and by **3.2.1** $R/I \oplus N \overset{F}{\leftrightarrow} R$. The converse proof is similar to that of **3.2.1**.

3.2.4 THEOREM   *Every family of rings contains a ring R with $\mathfrak{A}(R) \subseteq R^2$.*

Proof   Let R be a ring generated by a set of elements $\{x_i \mid i \in I\}$, I an index set. Let N be the null ring whose additive group is free, with generators $\{y_i \mid i \in I\}$. Then, by **3.1.4**, $R \overset{F}{\leftrightarrow} R \oplus N$. Let S denote the subring of $R \oplus N$ generated by $\{x_i + y_i \mid i \in I\}$. Since $S + N = R \oplus N$ and N is null, $S \overset{F}{\leftrightarrow} R \oplus N \overset{F}{\leftrightarrow} R$. Moreover, $S/S^2 \cong N$, and $Q = \mathfrak{A}(S)/(\mathfrak{A}(S) \cap S^2) \cong (\mathfrak{A}(S) + S^2)/S^2$. Hence, additively, Q is isomorphic to a subgroup of a free abelian group, so $Q = 0$ or Q is itself a null ring on a free abelian group. In either case $\mathfrak{A}(S)$ is the internal direct sum of $\mathfrak{A}(S) \cap S^2$ and another null ring K. Since $K \cap S^2 = 0$,

and $K \subseteq \mathfrak{A}(S)$ implies by (7) of **1.2.1** that K is an ideal in S, it follows by **3.2.3** that $S \overset{F}{\leftrightarrow} S/K$. Moreover,
$$\mathfrak{A}(S/K) = \mathfrak{A}(S)/K = ((\mathfrak{A}(S) \cap S^2) + K)/K \subseteq (S/K)^2.$$
Thus S/K has the required property.

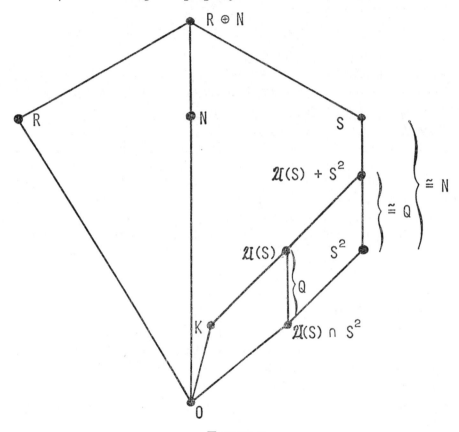

FIGURE 2

A ring R with $\mathfrak{A}(R) \subseteq R^2$ is called a *stem* ring. For finite rings we have the following

3.2.5 THEOREM *A finite ring R is a stem ring if and only if R has minimal order in its family.*

Proof Suppose $S \overset{F}{\leftrightarrow} R$. Then $[R:R^2 + \mathfrak{A}(R)] = [S:S^2 + \mathfrak{A}(S)]$, since $R/\mathfrak{A}(R) \simeq S/\mathfrak{A}(S)$. Thus, if R is a stem ring, then
$$|S| = [S:S^2 + \mathfrak{A}(S)] \cdot |S^2 + \mathfrak{A}(S)| = [R:R^2] \cdot |S^2 + \mathfrak{A}(S)| \geq$$

$[R:R^2] \cdot | S^2 | = [R:R^2] \cdot | R^2| = | R |$, with equality if and only if $\mathfrak{A}(S) \subseteq S^2$.

The vector space structure of an algebra permits many calculations which are not possible in an arbitrary ring, and thereby simplifies the study of algebra structure theory. Thus results which provide relations between classes of rings and algebras will significantly aid the study of rings. The great success of the Artin–Wedderburn theory is that it finds a strong relation of this type. For finite nilpotent rings the family classification provides one such relation, which is of considerable help in constructing rings of small order. We shall first show that for finite dimensional algebras the theory of families essentially reduces to isomorphism. For this result we shall require that the isomorphisms in the definition of family preserve scalar multiplication.

3.2.6 THEOREM *Suppose R and S are algebras over a field F in the same family of algebras over F. If R and S have the same finite dimension then R and S are isomorphic.*

Proof Define a map φ from R to S as follows. On a basis $r_1, \ldots, r_k$ for R^2, φ is the family isomorphism of R^2 onto S^2. Choose elements $r_{k+1}, \ldots, r_m$, which extend the basis to $R^2 + \mathfrak{A}(R)$, and elements $r_{m+1}, \ldots, r_n$ to extend the basis to R. $\varphi(r_1), \ldots, \varphi(r_k)$ are a basis for S^2. Let $s_{k+1}, \ldots, s_m$ extend this basis to $S^2 + \mathfrak{A}(S)$ and define $\varphi(r_i) = s_i, i = k + 1, \ldots, m$. For $m < i \leq n$ choose $s_i \in S$ so that $r_i + \mathfrak{A}(R) \to s_i + \mathfrak{A}(S)$ under the family isomorphism of $R/\mathfrak{A}(R)$ onto $S/\mathfrak{A}(s)$. Define $\varphi(r_i) = s_i$. This choice of s_i, $m < i \leq n$, completes a basis for S, for the set has the proper cardinality and spans S. Extend φ by linearity to all of R. The consistency of the family isomorphisms implies φ preserves products and hence is an isomorphism of R onto S.

3.2.7 COROLLARY *Let R and S be algebras over a field F in the same family of algebras over F, with dim $R <$ dim $S < \infty$. Then $S \cong R \oplus N$, where N is a null algebra over F.*

Proof Let N be the null algebra of dimension $\dim(S) - \dim(R)$ over F. By **3.1.4**, $R \overset{F}{\leftrightarrow} R \oplus N$. By hypothesis $S \overset{F}{\leftrightarrow} R$, so by the theorem S is isomorphic to $R \oplus N$.

If a ring R is in the same (ring) family as an algebra over a field F, then it is clearly necessary that $R/\mathfrak{A}(R)$ and R^2 be ring isomorphic to algebras over F. In case R is finite and F is a finite prime field, we now show the sufficiency of this condition for the existence of such an algebra.

3.2.8 THEOREM *Let R be a finite p-ring with* char $R^2 = p$, *or equivalently,* char $(R/\mathfrak{A}(R)) = p$. *Then there is a p-algebra A in the same family as R, and $|A| = |R|$.*

Proof The equivalence of the assumptions was noted in **1.2.5**. Let R have a basis $x_1, x_2, \ldots, x_n$, with char $x_i = p^{k_i}$, $1 \leq i \leq n$. Define the elementary abelian p-group A whose basis is the set of elements $\{y_{ij} \mid 1 \leq i \leq n, 0 \leq j < k_i\}$. There is a one-to-one correspondence α of R onto A given by

$$\alpha \left(\sum_{i=1}^{n} m_i x_i \right) = \sum_{i=1}^{n} \sum_{j=0}^{k_i-1} m_{ij} y_{ij},$$

where $m_i = \sum_j m_{ij} p^j$ is the integer m_i in p-ary form, $0 \leq m_{ij} < p$. Denote the inverse correspondence to α by ρ.

Multiplication is defined on A by $a_1 a_2 = \alpha(\rho(a_1)\rho(a_2))$. We now show that the distributive law holds in A. It will then follow from the one-to-one correspondence that A is an associative p-algebra which is in the same family as R, and $|A| = |R|$.

Let $R' = \{r \in R \mid pr = 0\}$. Then, for $r_1, r_2 \in R'$,

$$\alpha(r_1 + r_2) = \alpha(r_1) + \alpha(r_2).$$

Choose $a_i \in A$ and let $r_i = \rho(a_i)$, $i = 1, 2, 3$. Since R^2 has characteristic p, we have $R^2 \subseteq R'$, so

$$a_1 a_2 + a_1 a_3 = \alpha(r_1 r_2) + \alpha(r_1 r_3)$$
$$= \alpha(r_1 r_2 + r_1 r_3).$$

Next, observe that, for $a, b \in A$, $\rho(a + b) \equiv \rho(a) + \rho(b)$ (mod pR). Thus $a_1(a_2 + a_3) = \alpha(r_1(\rho(a_2 + a_3)))$
$$= \alpha(r_1[\rho(a_2) + \rho(a_3) + x]) \text{ for some } x \in pR$$
$$= \alpha(r_1[r_2 + r_3]) \text{ since } pR \subseteq \mathfrak{A}(R).$$

Thus distributivity in A is equivalent to distributivity in R. Thus A is an algebra.

Note that the correspondence α is an isomorphism of R^2 onto A^2, carries $\mathfrak{A}(R)$ onto $\mathfrak{A}(A)$, and induces a consistent (family) isomorphism of $R/\mathfrak{A}(R)$ onto $A/\mathfrak{A}(A)$. Thus $R \overset{F}{\leftrightarrow} A$, and the proof of **3.2.8** is complete.

We now wish, conversely, to construct all p-rings in the same family with a given finite dimensional p-algebra. To do so let us continue to study the p-algebra A constructed from a p-ring R in the proof of **3.2.8**.

Since $pR \subseteq \mathfrak{A}(R)$ and char $R^2 = p$, we can use the correspondence α to decompose A as a direct sum $A \cong B \oplus N$, where

$$B = \langle y_{i\,0}, y_{i\ k_i-1} \mid 1 \leq i \leq n \rangle$$

and $N = \langle y_{ij} \mid 1 \leq i \leq n, 1 \leq j < k_i - 1 \rangle$. By **3.1.4**, $B \overset{F}{\hookrightarrow} A$.

Denote the additive group of R by M. Reorder the basis $\{x_1, \ldots, x_n\}$ of M so that

(1) M has a basis $\{x_1, \ldots, x_m, \ldots, x_n\}$ with char $x_i > p$ if $1 \leq i \leq m$, and char $x_i = p$ if $m < i \leq n$.

It then follows that dim $B = m + n$, and if we define

$$b_i = y_{i\,0}, 1 \leq i \leq n, b_{n+i} = y_{i\ k_i-1}, 1 \leq i \leq m,$$

then B has a basis $\mathfrak{B} = \{b_1, \ldots, b_{n+m}\}$ such that

(2) $B^2 \subset \langle b_{m+1}, \ldots, b_{n+m} \rangle$ and $\mathfrak{U}(B) \supset \langle b_{n+1}, \ldots, b_{n+m} \rangle$.

If, on the other hand, we start with M, B, and any basis $\mathfrak{B}'$ for B satisfying (2), then we clearly have the following

3.2.9 LEMMA *Let M be an abelian p-group satisfying (1), and B a p-algebra with a basis $\mathfrak{B}' = \{b_1', \ldots, b_{n+m}'\}$ satisfying (2). The correspondence $\rho(b_i') = x_i, 1 \leq i \leq n$, $\rho(b_{n+i}') = p^{k_i-1}x_i, 1 \leq i \leq m$, then defines a ring R' whose additive group is M, such that $R' \overset{F}{\leftrightarrow} B$. Multiplication is defined in R' by $x_i x_j = \rho(b_i b_j)$, with extension by distributivity to all of R'.*

Finally, we must determine when different bases of B which satisfy (2) determine isomorphic rings under the construction **3.2.9**. To do so define $\mathfrak{S}$ to be the set of all bases for B which satisfy (2), and $\mathfrak{T}$ to be the automorphism group of B, regarded as a permutation group on $\mathfrak{S}$. Since the definition of R' in **3.2.9** depends only on the multiplicative behaviour of the basis $\mathfrak{B}'$, the ring R' depends only on the orbit of $\mathfrak{B}'$ under the action of $\mathfrak{T}$. On the other hand, any ring R'' isomorphic to R' may be regarded as having additive group M, in which case it differs from R only by a change of basis of M which preserves (1). The group of changes of basis of M preserving (1) induces, via the correspondence $\alpha = \rho^{-1}$, a group $\mathfrak{U}$ of permutations on $\mathfrak{S}$. Thus $\mathfrak{S}$ is divided into orbits under the action of $\mathfrak{T}$ and $\mathfrak{U}$ together, and these orbits correspond to mutually non-isomorphic rings. We now summarize these results.

3.2.10 THEOREM *Suppose M is an abelian p-group satisfying (1), B is a p-algebra and $\mathfrak{S}$ is the set of bases of B satisfying (2), and $\mathfrak{T}$ and $\mathfrak{U}$ are the groups of permutations of $\mathfrak{S}$ defined above. Then there is a one-to-one correspondence between the orbits of $\mathfrak{S}$ under the action of $\mathfrak{T}$ and $\mathfrak{U}$ together, and the rings R with additive group M in the same family with B.*

For computational purposes it is convenient to reformulate the definitions of $\mathfrak{S}$, $\mathfrak{T}$, and $\mathfrak{U}$ so that all actions are expressed in terms of a fixed basis $\mathfrak{B}_0$ for B. To do so let us express $\mathfrak{B} \in \mathfrak{S}$ as a matrix of column vectors, where column i of $\mathfrak{B}$ expresses the ith basis member of $\mathfrak{B}$ as a linear combination of the basis members of $\mathfrak{B}_0$. Further let us write an automorphism $T \in \mathfrak{T}$ as a matrix whose i^{th} column is the image of the ith basis member in $\mathfrak{B}_0$, expressed in terms of the basis $\mathfrak{B}_0$. Then the set $\mathfrak{S}$ is closed under multiplication on the left by $\mathfrak{T}$ and its elements are permuted regularly by $\mathfrak{T}$. In a similar way let us consider $U \in \mathfrak{U}$ as a matrix in terms of the basis $\mathfrak{B}_0$. Then $\mathfrak{S}$ is closed under multiplication on the right by $\mathfrak{U}$, and since the correspondence α is one-to-one, $\mathfrak{U}$ acts regularly on $\mathfrak{S}$. Thus $\mathfrak{S}$ is decomposed into double cosets of the form $\mathfrak{T} \mathfrak{B} \mathfrak{U}$, $\mathfrak{B} \in \mathfrak{S}$, and the double cosets correspond to the mutually non-isomorphic rings R with $R^+ = M$ and $R \overset{F}{\leftrightarrow} B$.

3 An example: the nilpotent rings of order p^3

From now on the family classification will be used as part of our standard notation for finite nilpotent p-rings. Families will be denoted by capital letters. A ring R will be denoted $X n_1 \ldots n_k y$, where R is in family X; R^+ has type $_p(n_1, \ldots, n_k)$; and y is a lower case letter specifying R within its family and additive group type. If the definition of either the family or the ring within the family depends on variable structure constants, these will be written as subscripts to the designators X and y, respectively. The designator y will sometimes be omitted when there is only one ring in the family with the given additive group type. In particular, y will always be omitted for null rings, cyclic rings, and p-algebras.

3.3.1 EXAMPLE *The nilpotent rings of orders p and p^2 are:*
$Z1$: *Null ring of order p.*
$Z11$: *Null ring of type $_p(1, 1)$.*
$Z2$: *Cyclic null ring of order p^2.*
$A11$: *The p-algebra with basis $\{a, a^2\}$, $a^3 = 0$.*
$A2$: *The cyclic ring of order p^2 generated by a, with $a^2 = pa$.*

We now turn to the family classification of the nilpotent rings of order p^3. By **3.2.6** the p-algebras of dimension 3 all lie in different families. We then have the following list, as given in **2.3.6**.

3.3.2 *A nilpotent p-algebra A of dimension 3 has a basis $\{a, b, c\}$, with $c \in \mathfrak{A}(A)$, which satisfies one of the following conditions. These cases are mutually non-isomorphic.*

$Z111$: *Null algebra.*

$A111$: $a^2 = b$, $a^3 = 0$: $A11 \oplus Z1$.

$B111$: $a^2 = b^2 = 0$, $ab = -ba = c$.

$C_\gamma 111$: $a^2 = c$, $b^2 = \gamma c$, $ab = ba = 0$, *and $\gamma = 1$ or γ is a fixed non-square mod p, $p \neq 2$.*

$D_\phi 111$: $a^2 = ab = c$, $ba - 0$, $b^2 = \varphi c$, $0 \leq \varphi < p$.

$E111$: $a^2 = b$, $a^3 = c$, $a^4 = 0$.

There are, by **2.3.2**, three non-isomorphic cyclic nilpotent rings of order p^3. If $R = \langle a \rangle$ with char $a = p^3$, $a^2 = 0$, then R is in family Z, while if $a^2 = p^2 a$, then R is in family A. If $a^2 = pa$, then char $R^2 = p^2$, so R cannot be in any of the families $Z - E$. Thus we have

3.3.3 *A cyclic nilpotent ring of order p^3 contains an element a, char $a = p^3$, such that one of the following conditions holds:*

$Z3$: $a^2 = 0$.

$A3$: $a^2 = p^2 a$.

$F3$: $a^2 = pa$.

It remains for us to classify the nilpotent rings of type $_p(2, 1)$. If R is such a ring, then obviously char $R^2 = p$, so by **3.2.8** R is in one of the families $Z - E$. If R is in family Z then R is null. We shall now construct the nilpotent rings of type $_p(2, 1)$ in families $A - E$ by the method of Theorem **3.2.10**, using the matrix notation given after the theorem, with all matrix entries from $GF(p)$ and all numerical computations performed in $GF(p)$. The rings constructed will all have a basis $\{x, y\}$ with char $x = p^2$, char $y = p$. Because the computations are slightly simpler, we shall begin with family B and consider family A last.

Family B It is easy to check that $S \in \mathfrak{S}$, $T \in \mathfrak{T}$, and $U \in \mathfrak{U}$ have the form

$$S = \begin{pmatrix} s_1 & s_4 & 0 \\ s_2 & s_5 & 0 \\ s_3 & s_6 & s_7 \end{pmatrix}, \quad T = \left(\begin{array}{cc|c} T_1 & & 0 \\ & & 0 \\ \hline t_2 & t_3 & t_4 \end{array} \right) \quad U = \begin{pmatrix} u_1 & 0 & 0 \\ u_2 & u_4 & 0 \\ u_3 & u_5 & u_1 \end{pmatrix}$$

where $\begin{pmatrix} s_1 & s_4 \\ s_2 & s_5 \end{pmatrix}$ and T_1 are non-singular 2×2 blocks, $t_4 = $ determinant (T_1), and u_1, u_4, and s_7 are not 0. Clearly there is only one double coset $\mathfrak{T} S \mathfrak{U}$ in $\mathfrak{S}$, and it is represented by the identity matrix I. Thus there is only one ring:

$$B21: xy = -yx = px, \ x^2 = y^2 = 0.$$

Families C_γ, $\gamma = 1$ or γ is a fixed non-square mod p　$\mathfrak{S}$ and $\mathfrak{U}$ have the same structure as for family B. To find the structure of the automorphism group $\mathfrak{T}$ of $C_\gamma 111$, consider an arbitrary automorphism

$$a' = t_1 a + t_2 b + t_3 c,$$
$$b' = t_4 a + t_5 b + t_6 c,$$
$$c' = a'^2,$$

where $\{a, b, c\}$ is a basis for $C_\gamma 111$ such that $a^2 = c$, $b^2 = \gamma c$, $ab = ba = 0$. The conditions $b'^2 = \gamma a'^2 = c' \neq 0$ and $a'b' = b'a' = 0$ are equivalent to

$$t_4^2 + t_5^2 \gamma = \gamma(t_1^2 + t_2^2 \gamma) \neq 0$$

and
$$t_1 t_4 + t_2 t_5 \gamma = 0$$

These conditions imply $t_4 = \varepsilon \gamma t_2$ for $\varepsilon = \pm 1$, $t_5 = -\varepsilon t_1$, and $t_1^2 + \gamma t_2^2 \neq 0$. Thus $T \in \mathfrak{T}$ has the form

$$T = \begin{pmatrix} t_1 & \varepsilon \gamma t_2 & 0 \\ t_2 & -\varepsilon t_1 & 0 \\ t_3 & t_6 & t_1^2 + \gamma t_2^2 \end{pmatrix}.$$

The first ring in family C_γ corresponds to the coset in $\mathfrak{S}$ containing $S_1 = I$. This ring is:

$$C_\gamma 21a \colon x^2 = px, \ y^2 = \gamma px, \ xy = yx = 0.$$

It is easy to check that
$$|\mathfrak{S}| = (p^2 - 1)\, p^3 (p - 1)^2, \ |\mathfrak{U}| = p^3(p - 1)^2, \ |\mathfrak{T} \cap \mathfrak{U}| = 2p^2$$
if $p \neq 2$, and $|\mathfrak{T} \cap \mathfrak{U}| = 4$ if $p = 2$. In the case when $-\gamma$ is not a square in $GF(p)$ then $|\mathfrak{T}| = 2p^2(p^2 - 1)$, so
$$|\mathfrak{T} I \mathfrak{U}| = |\mathfrak{T}| \cdot |\mathfrak{U}| / |\mathfrak{T} \cap \mathfrak{U}| = |\mathfrak{S}|,$$
so in this case $C_\gamma 21a$ is the only ring. We now suppose $-\gamma$ is a square in $GF(p)$. In this case $|\mathfrak{T}| = 2p^2(p - 1)^2$, so $|\mathfrak{T} I \mathfrak{U}|$ is $p^3(p - 1)^4$ if $p \neq 2$ and is 16 if $p = 2$. Thus $\mathfrak{T} I \mathfrak{U}$ contains precisely those $S \in \mathfrak{S}$ for which $s_4^2 + s_5^2 \gamma \neq 0$. Hence a representative of another coset is

$$S_2 = \begin{pmatrix} 1 & \gamma & 0 \\ 0 & \psi & 0 \\ 0 & 0 & 1 \end{pmatrix}$$

where $\psi^2 = -\gamma$. This gives the ring

$$C_\gamma 21b \colon x^2 = px, \ xy = yx = \gamma px, \ y^2 = 0.$$

Finally, it can be shown that there are no further cosets in $\mathfrak{S}$ by computation of $|\mathfrak{T} S_2 \mathfrak{U}|$.

Families D_ϕ, $0 \leq \varphi < p$　$\mathfrak{S}$ and $\mathfrak{U}$ again have the same structure as for family B. To find the structure of $T \in \mathfrak{T}$ consider an automorphism of $D_\phi 111$ given by

$$a' = t_1 a + t_2 b + t_3 c,$$
$$b' = t_4 a + t_5 b + t_6 c,$$
$$c' = a'^2,$$

where $\{a, b, c\}$ is a basis for $D_\phi 111$ such that $a^2 = ab = c$, $ba = 0$, and $b^2 = \varphi c$. The computation in the proof of **2.3.6** gives $t_4 = -t_2 \varphi$ and $t_5 = t_1 + t_2$. $c' = a'^2 \neq 0$ requires $t_1^2 + t_1 t_2 + t_2^2 \varphi \neq 0$. Hence we have

$$T = \begin{pmatrix} t_1 & -t_2\varphi & 0 \\ t_2 & t_1 + t_2 & 0 \\ t_3 & t_6 & t_1^2 + t_1 t_2 + t_2^2\,\varphi \end{pmatrix}$$

First, as usual, we have the ring corresponding to the coset containing the identity I:

$$D_\varphi 21a \colon x^2 = xy = px, \, yx = 0, \, y^2 = \varphi px.$$

If $1 - 4\varphi$ is not a square in $GF(p)$, then $|\mathfrak{T}| = p^2(p^2 - 1)$, and $|\mathfrak{T} \cap \mathfrak{U}| = p^2$, so $|\mathfrak{T} I \mathfrak{U}| = |\mathfrak{T}| \cdot |\mathfrak{U}|/|\mathfrak{T} \cap \mathfrak{U}| = |\mathfrak{S}|$. Thus $D_\phi 21a$ is the only ring of type $_p(2, 1)$ in its family. Similarly, if $p = 2$ and $\varphi = 1$ then $|\mathfrak{T} I \mathfrak{U}| = |\mathfrak{S}|$, so there is only one ring.

Henceforth we assume that either $\varphi = 0$, or else $\varphi \neq 0$, $p \neq 2$, and $1 - 4\varphi = \psi^2$ for some $\psi \in GF(p)$.

First we consider the case $\varphi = 0$. Then all $S \in \mathfrak{T} I \mathfrak{U}$ have $s_4 = 0$, and, since $|\mathfrak{T} I \mathfrak{U}| = p^3(p - 1)^3$, all S with $s_4 = 0$ are in $\mathfrak{T} I \mathfrak{U}$. Thus we have a second coset, represented by

$$S_2 = \begin{pmatrix} 0 & 1 & 0 \\ 1 & 0 & 0 \\ 0 & 0 & 1 \end{pmatrix},$$

and a corresponding ring

$$D_0 21b \colon x^2 = xy = 0, \, yx = y^2 = px.$$

All $S \in \mathfrak{T} S_2 \mathfrak{U}$ satisfy $s_4 + s_5 \neq 0$. Hence there is a third coset, represented by

$$S_3 = \begin{pmatrix} 1 & 1 & 0 \\ 0 & -1 & 0 \\ 0 & 0 & 1 \end{pmatrix},$$

and a corresponding ring

$$D_0 21c \colon x^2 = yx = px, \, xy = y^2 = 0.$$

A computation of the orders of the cosets show that these exhaust $\mathfrak{S}$. Note that rings $D_0 21a$ and $D_0 21c$ are anti-isomorphic.

We now suppose $\varphi \neq 0, p \neq 2$, and $1 - 4\varphi = \psi^2$ for some $\psi \in GF(p)$. If, then, $S \in \mathfrak{T} I \mathfrak{U}$, it follows that

$$s_4^2 + s_4 s_5 + s_5^2\,\varphi \neq 0. \tag{1}$$

Since $|\, \mathfrak{T}\, I\, \mathfrak{U}\, | = p^3(p-1)^4$, moreover, all S satisfying (1) are in $\mathfrak{T}\, I\, \mathfrak{U}$. Instead of directly determining the cosets in $\mathfrak{S}$ for which (1) fails, let us for simplicity note that condition (1) for $S \in \mathfrak{S}$ is equivalent to $y^2 \neq 0$ in the ring of type $_p(2, 1)$ corresponding to S. Further, a nilpotent ring R of order p^3 is in one of the families D_ϕ if and only if R is not commutative and there exists $x \in R$ with $x^2 \neq 0$. Hence it is sufficient for us to characterize and classify by family the nilpotent rings R with a basis $\{x, y\}$ such that char $x = p^2$, char $y = p$, $x^2 = px$, $y^2 = 0$, $xy = \alpha px$, and $yx = \beta px$, some $\alpha, \beta \in GF(p)$, $\alpha \neq \beta$.

First let us note that $\alpha = 0$ or $\beta = 0$ implies $R \in D_0$. Hence suppose $\alpha \neq 0$ and $\beta \neq 0$. Replacing y by $\alpha^{-1}y$ we obtain $\alpha = 1$, so $\beta \neq 1$. It is easy to check that β cannot be changed within an isomorphism class, given $\alpha = 1$. Hence we have $p - 2$ distinct rings R_β. Now define $u = (1 - \beta)x$, $v = y - \beta x$, and $w = (1 - \beta)^2 px$. Then $u^2 = uv = w$, $vu = 0$, and $v^2 = [-\beta/(1 - \beta)^2]w$, so R is in family D_ϕ, where $\varphi = -\beta/(1 - \beta)^2$. Solving for β we obtain $\beta = 1 - (1 \pm \psi)/2\varphi$ where $\psi^2 = 1 - 4\varphi$. Thus we have

If $4\varphi \equiv 1 \pmod{p}$ there is one ring:

$$D_\phi 21b\colon x^2 = xy = px,\ y^2 = 0,\ yx = [1 - 1/(2\varphi)]px.$$

If $4\varphi \not\equiv 1 \pmod{p}$ there are two rings:

$$D_\phi 21b\colon x^2 = xy = px,\ y^2 = 0,\ yx = [1 - (1 + \psi)/2\varphi]px.$$
$$D_\phi 21c\colon x^2 = xy = px,\ y^2 = 0,\ yx = [1 - (1 - \psi)/2\varphi]px.$$

Family E   If R is a ring in family E then $|\, R^2\, | = p^2$, so if $|\, R\, | = p^3$ then R is a power ring by **2.3.1**. If R has type $_p(2, 1)$ it follows that a basis $\{x, y\}$ for R may be chosen so that $x^2 = y$, $x^3 = xy = yx = \alpha px$ for some $\alpha \in GF(p)$. Rings R in family E have $R^3 \neq 0$, so $\alpha \neq 0$. Let $x' = \varphi x$, $y' = \varphi^2 y$ for $\varphi \in GF(p)$, $\varphi \neq 0$. Then $x'^2 = y'$, $x'^3 = \varphi^2 \alpha px'$. Hence we may choose $\{x, y\}$ so that either $\alpha = 1$ or α is a fixed non-square in $GF(p)$. Thus we have the rings:

$E21a_\psi\colon x^2 = y$, $x^3 = \psi px$, $x^4 = 0$, $\psi = 1$ or $p \neq 2$ *and* ψ *is a fixed non-square* (mod p).

Family A $\mathfrak{U}$ has the same structure as for family B. Let algebra $A111$ have a basis $\{a, b, c\}$ with $a^2 = b$, all other products 0. Then $S \in \mathfrak{S}$ and $T \in \mathfrak{T}$ have the structure

$$S = \begin{pmatrix} s_1 & s_4 & 0 \\ s_2 & s_5 & s_7 \\ s_3 & s_6 & s_8 \end{pmatrix} \quad T = \begin{pmatrix} t_1 & 0 & 0 \\ t_2 & t_1^2 & t_4 \\ t_3 & 0 & t_5 \end{pmatrix}$$

where S is nonsingular, t_1 and t_5 are not 0, and either $s_4 = 0$, or else

$s_4 \neq 0$, $s_7 \neq 0$, and $s_8 = 0$. Let

$$\mathfrak{S}_1 = \{S \in \mathfrak{S} \mid s_4 = 0\} \text{ and } \mathfrak{S}_2 = \{S \in \mathfrak{S} \mid s_4 \neq 0\}.$$

It is easy to check that elements of $\mathfrak{S}_1$ and $\mathfrak{S}_2$ must lie in different cosets, so we shall consider two cases.

First suppose $s_4 = 0$. The conditions $s_8 = 0$ and $s_8 \neq 0$ are preserved under left multiplication by $T \in \mathfrak{T}$ and right multiplication by $U \in \mathfrak{U}$, so there are at least two cosets, represented by

$$\begin{pmatrix} 1 & 0 & 0 \\ 0 & 1 & 0 \\ 0 & 0 & 1 \end{pmatrix} \quad \text{and} \quad \begin{pmatrix} 1 & 0 & 0 \\ 0 & 0 & 1 \\ 0 & 1 & 0 \end{pmatrix},$$

and two corresponding rings:

$$A21a\colon x^2 = y,\ x^3 = 0.$$
$$A21b\colon x^2 = px,\ xy = yx = y^2 = 0.$$

Computation of their orders shows that these two cosets exhaust $\mathfrak{S}_1$. Note that $A21b$ is isomorphic to $A2 \oplus Z1$.

Second suppose $s_4 \neq 0$, $s_7 \neq 0$, and $s_8 = 0$. There is only one coset in $\mathfrak{S}_2$, represented by

$$\begin{pmatrix} 0 & 1 & 0 \\ 0 & 0 & 1 \\ 1 & 0 & 0 \end{pmatrix}.$$

The corresponding ring is

$$A21c\colon y^2 = px,\ x^2 = xy = yx = 0.$$

We now summarize the above computations.

3.3.4 *A nilpotent ring of type $_p(2, 1)$ has a basis $\{x, y\}$, with* char $x = p^2$, char $y = p$, *which satisfies one of the following conditions. These cases are mutually non-isomorphic. For $p \neq 2$, μ denotes a fixed non-square* (mod p).

$Z21$: $x^2 = xy = yx = y^2 = 0$.   *Null ring.*

$A21a$: $x^2 = y,\ x^3 = 0$.   *Power ring.*

$A21b$: $x^2 = px,\ xy = yx = y^2 = 0$.   *Isomorphic to $A2 \oplus Z1$.*

$A21c$: $y^2 = px,\ x^2 = xy = yx = 0$.

$B21$: $xy = -yx = px,\ x^2 = y^2 = 0$.

$C_\gamma 21a$: $x^2 = px,\ y^2 = \gamma px,\ xy = yx = 0,\ \gamma = 1,$ *or* $p \neq 2$ *and* $\gamma = \mu$.

$C_\gamma 21b$: $\gamma = 1$ *or* $\gamma = \mu$, *and* $-\gamma$ *is a square* (mod p). $x^2 = px$,
          $xy = yx = \gamma px,\ y^2 = 0$.

$D_\varphi 21a$: $x^2 = xy = px,\ yx = 0,\ y^2 = \varphi px,\ \varphi$ *is an integer,* $0 \leq \varphi < p$.

$D_0 21b$: $x^2 = xy = 0,\ yx = y^2 = px$.

$D_0 21c$: $x^2 = yx = px,\ xy = y^2 = 0$.   *Anti-isomorph of $D_0 21a$.*

$D_\phi 21b$: $1 \leq \varphi < p$, $1 - 4\varphi \equiv \psi^2$ (mod p), *some integer* ψ, $p \neq 2$:
$x^2 = xy = px$, $y^2 = 0$, $yx = [1 - (1 + \psi)/2\varphi]px$.

$D_\phi 21c$: φ, ψ, p *as for* $D_\phi 21b$, *and* $\psi \not\equiv 0$ (mod p): $x^2 = xy = px$,
$y^2 = 0$, $yx = [1 - (1 - \psi)/2\varphi]px$.

$E21a_\psi$: $x^2 = y$, $x^3 = \psi px$, $x^4 = 0$, $\psi = 1$, *or* $p \neq 2$ *and* $\psi = \mu$.

4 Nilpotent rings with chain conditions

This section discusses some properties of nilpotent rings with descending chain condition (d.c.c.) or ascending chain condition (a.c.c.) on two-sided ideals. The principal result is

3.4.1 **THEOREM** *A nilpotent ring which satisfies the descending chain condition for ideals is in the same family as a finite nilpotent ring.*

The proof requires the following three lemmas.

3.4.2 **LEMMA** *If a nilpotent ring R satisfies the descending chain condition for ideals, then R^+ satisfies the descending chain condition for subgroups.*

Proof We use induction on $e = \exp(R)$. For $e = 2$ the result follows from (7) of **1.2.1**. Suppose $e > 2$. Again by (7) of **1.2.1** the subgroups of $(R^{e-1})^+$ satisfy the d.c.c. By induction hypothesis $(R/R^{e-1})^+$ satisfies the d.c.c. for subgroups. Thus we need only show for B, a subgroup of an abelian group A, that if B and A/B satisfy the d.c.c. for subgroups, then A does also. Suppose $K_1 \supset K_2 \supset \ldots$ were an infinite strictly descending chain of subgroups of A. Since B has d.c.c., there is an index n such that $K_n \cap B = K_i \cap B$, all $i \geq n$. Thus

$$K_n/(K_n \cap B) \supset K_{n+1}/(K_{n+1} \cap B) \supset \ldots$$

would be strictly descending. But $K_j/(K_j \cap B) \cong (K_j + B)/B$. This would construct a strictly descending infinite chain of subgroups in A/B, which is impossible.

3.4.3 **LEMMA** *An abelian group satisfies the descending chain condition for subgroups if and only if it is isomorphic to a finite direct sum of groups $C(p_i^{n_i})$, $1 \leq n_i \leq \infty$.*

Proof Let G be an abelian group satisfying the d.c.c. for subgroups. Then G obviously contains no element of infinite order, and so G is isomorphic to a finite direct sum of p-groups. Hence without loss of generality we can assume that G itself is a p-group. Since G satisfies the d.c.c., the subgroup $A = \{x \in G \mid px = 0\}$ is finite. If $A = 0$ the result is trivial. We proceed by induction on $|A|$. If G contains a subgroup $B \neq 0$ with $pB = B$, then there is an infinite set of elements b_1, b_2, ... in B such that $b_1 \neq 0$, $pb_1 = 0$, and $pb_{i+1} = b_i$, $i \geq 1$. This set generates a quasi-cyclic subgroup $Q \cong C(p^\infty)$, which must be a direct summand of G. Let G' be a subgroup of G such that $G \cong Q \oplus G'$. By induction hypothesis G' is isomorphic to a finite direct sum of groups $C(p_i^{n_i})$, $1 \leq n_i \leq \infty$, whence G is also. If, on the other hand, G contains no subgroup $B \neq 0$ with $pB = B$, then, since the chain of subgroups $G \supseteq pG \supseteq p^2G \supseteq \ldots$ must terminate, $p^tG = 0$ for some natural number t. It follows that G is isomorphic to a direct sum of cyclic subgroups. Since G satisfies the d.c.c., this sum is finite. Thus, in each case, if G satisfies the d.c.c. then G is isomorphic to a finite direct sum of groups $C(p_i^{n_i})$, $1 \leq n_i \leq \infty$. The converse is clear.

3.4.4 LEMMA *If x is an element of infinite height in a p-ring R, then $x \in \mathfrak{A}(R)$.*

Proof Choose $y \in R$. Since x has infinite height, there exists $z \in R$ such that $x = (\text{char } y)z$. Thus $xy = z(\text{char } y)y = 0$. Similarly $yx = 0$, so $x \in \mathfrak{A}(R)$.

Proof of **3.4.1** Let R be an infinite nilpotent ring with the d.c.c. for ideals. By **3.4.2** and **3.4.3** R is a torsion ring, so by **1.5.3** we may suppose that R is a p-ring. Let R^+ be isomorphic to a finite direct sum of rings $C(p^{n_i})$, $1 \leq n_i \leq \infty$. Let n be the largest finite n_i, and let $S = \{x \in R \mid p^nx = 0\}$. Then S is a finite nilpotent ring, and, by **3.4.4** and **3.2.1**, is in the same family as R.

Theorem **3.4.1** was obtained by T. Szele [3] in 1955. Szele calls the finite subring S the "kernel" of the p-ring R. Let us note that a finite ring S occurs as the kernel of only finitely many non-isomorphic infinite nilpotent p-rings R with d.c.c. Any such ring R can be constructed by selecting an independent set of elements $\{x_1, \ldots, x_k\}$ from S such that $x_i \in \mathfrak{A}(S)$ and char $x_i =$ char S, $1 \leq i \leq k$. To obtain R replace each cyclic subgroup $\{x_i\}$ by a quasi-cyclic p-group $C(p^\infty)$.

Historical remark A section devoted to nilpotent rings with d.c.c. appears in the classic 1939 paper of Ch. Hopkins [2], who proves that if R is a nilpotent ring with the d.c.c. for left ideals, then R satisfies the d.c.c. for right ideals and for subrings, and R^+ is a torsion group. The results of Szele have been extended by R. Wiegandt [1], who obtains a connection between transfinitely nilpotent topological rings and inverse systems of finite nilpotent rings. Lemma **3.4.2** was first

D

obtained by S. A. Jennings [2] in 1944, who proves the result in a much more general context. Lemma **3.4.3** is due to A. G. Kurosh [1].

We conclude this section with some remarks about nilpotent rings satisfying the ascending chain condition (a.c.c.). It is well-known that an abelian group satisfies the a.c.c. for subgroups if and only if it is finitely generated. The next result, which clearly could be proved by an argument dual to that for **3.4.2**, will instead be obtained by the use of this remark.

3.4.5 LEMMA *Let R be a nilpotent ring with ascending chain condition on ideals. Then R^+ is a finitely generated abelian group.*

Proof Since every subgroup of $(R/R^2)^+$ is an ideal of R, it follows that $(R/R^2)^+$ is finitely generated. Choose $x_1, \ldots, x_n$ in R whose images mod R^2 generate R/R^2. Let $e = \exp R$. Then R^+ is generated by the no more than

$$n + n^2 + \ldots + n^{e-1}$$

distinct products of $e - 1$ or fewer generators x_i.

3.4.6 COROLLARY *If R is a nilpotent ring then the ascending chain conditions on additive subgroups, subrings, right ideals, and two-sided ideals are equivalent.*

3.4.7 COROLLARY *A finitely generated nilpotent ring has only finitely many subrings of a given finite index. A subring of a finitely generated nilpotent ring is finitely generated.*

It may be of interest to note that **3.4.7** holds without assuming nilpotence, although the general proof is much more difficult. This result is due to J. Lewin [1].

5 Capability; the construction of families

Given a ring S, this section describes a method of constructing all the families of rings R for which $S \cong R/\mathfrak{A}(R)$. This method is the study of a certain quotient ring (the *Schur Multiplier*) in a free ring.

A ring S is called *capable* if there exists a ring R for which $S \cong R/\mathfrak{A}(R)$. We shall obtain a necessary and sufficient condition that a ring be capable.

Problem: *For a ring S find all families of rings R for which $S \cong R/\mathfrak{A}(R)$.*

Let P be a free ring on indeterminates $\{x_i \mid i \in I\}$, I an index set. Let K be an ideal of P for which $S \cong P/K$. Suppose R is a ring for which $S \cong R/\mathfrak{A}(R)$. Let $\bar{\gamma}$ be the composite isomorphism $P/K \to S \to R/\mathfrak{A}(R)$. Choose $y_i \in R$, $i \in I$, so that $\bar{\gamma}(x_i + K) = y_i + \mathfrak{A}(R)$. Let $R_1 = \langle y_i \mid i \in I \rangle$ and let $\gamma\colon P \to R_1$ be the homomorphism defined by $\gamma(x_i) = y_i$. By

3.2.1, $R_1 \overset{F}{\hookrightarrow} R$. Let M be the kernel of γ. Certainly $M \subseteq K$. Moreover, $K = \gamma^{-1}(\mathfrak{A}(R_1))$. For $x \in K$ implies $\gamma(x) \in \mathfrak{A}(R) \cap R_1 \subseteq \mathfrak{A}(R_1)$, while $\gamma(x) \in \mathfrak{A}(R_1)$ implies, by **3.2.2**, that $\gamma(x) \in \mathfrak{A}(R) \cap R_1$, so

$$\bar{\gamma}(x + K) \in \mathfrak{A}(R),$$

and thus $x \in K$. Hence $K/M = \mathfrak{A}(P/M)$. Let $N_0 = KP + PK$. From $K/M = \mathfrak{A}(P/M)$ and $K \supseteq M \supseteq N_0$ follows $K/N_0 = \mathfrak{A}(P/N_0)$.

On the other hand, from $K/N_0 = \mathfrak{A}(P/N_0)$ follows

$$S \cong (P/N_0)/\mathfrak{A}(P/N_0).$$

Thus we have proved the following:

3.5.1 **THEOREM** *Let K be an ideal of a free ring P for which $S \cong P/K$. Then S is capable if and only if $K/N_0 = \mathfrak{A}(P/N_0)$, where $N_0 = PK + KP$.*

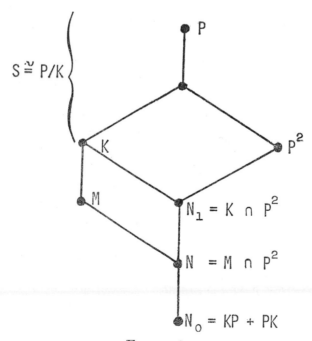

FIGURE 3

Next define $N_1 = K \cap P^2$ and $N = M \cap P^2$. Then
$$M/N \cap P^2/N = 0,$$
so, by **3.2.3**, $P/N \overset{F}{\leftrightarrow} (P/N)/(M/N) \cong P/M \cong R_1 \overset{F}{\leftrightarrow} R$. $K/N = \mathfrak{A}(P/N)$
follows. Hence we have proved:

3.5.2 THEOREM *Every family of rings R with $S \cong R/\mathfrak{A}(R)$ has a representative among the rings P/N such that*
(1) $N_0 \subseteq N \subseteq N_1$ *and*
(2) $\mathfrak{A}(P/N) = K/N$.

We shall call an ideal N of P which satisfies (1) and (2) of **3.5.2** a *capable ideal*. The family which contains P/N_0 is called the *maximal family* for S. Note that, for $N_0 \subseteq N \subseteq N_1$, $K/N \subseteq \mathfrak{A}(P/N)$ is trivial, so to check (2) requires showing only the reverse containment. Further, since $N_1 \subseteq K$ and $K/N_0 \subseteq \mathfrak{A}(P/N_0)$, it follows by (7) of **1.2.1** that every additive subgroup N of P^+, $N_0 \subseteq N \subseteq N_1$, must be an ideal of P.

Next we investigate the question of when two capable ideals, N and N^*, induce the same family. Let $T = P/N$, $T^* = P/N^*$. Then, by the definition of family and $T/\mathfrak{A}(T) = T^*/\mathfrak{A}(T^*) = P/K$, $T \overset{F}{\leftrightarrow} T^*$ if and only if there is an automorphism φ of P/K and a consistent isomorphism ψ between P^2/N and P^2/N^*. But $N_0 = PK + KP$, and hence φ induces a well defined automorphism $\bar\varphi$ of P^2/N_0, which satisfies the family consistency criterion. Thus ψ is a restriction of $\bar\varphi$. Thus

3.5.3 THEOREM *Two capable ideals, N and N^*, correspond to the same family if and only if there is an automorphism φ of P/K whose induced automorphism $\bar\varphi$ of P^2/N_0 takes N onto N^*.*

The rest of this section is devoted to examples illustrating the above results.

3.5.4 *If N is a finite null ring, then N is capable.*

Proof Let N^+ be isomorphic to a direct sum of cyclic p-groups $C(p_i^{n_i})$. Define a ring R as the direct sum of cyclic rings $\langle x_i \rangle$ where char $x_i = p_i^{2n_i}$ and $x_i^2 = p_i^{n_i} x_i$. Then $N \cong R/\mathfrak{A}(R)$.

Let us note that the group analogue of **3.5.4** does not hold. A finite abelian p-group is capable if and only if its largest two invariants are equal (P. Hall [3]).

3.5.5 *Every null algebra is capable.*

Proof Let N be a null algebra of dimension α over a field F, α any cardinal. Then $N \cong A/\mathfrak{A}(A)$, where A is the (restricted) direct sum of α copies of the power algebra of dimension 2 over F.

3.5.6 *The null ring whose additive group is the Prüfer quasi-cyclic p-group is not capable.*

Proof Suppose the contradiction. Then there is a ring R containing elements x_i, $1 \leq i < \infty$, with $x_i - px_{i+1} \in \mathfrak{A}(R)$, $1 \leq i$, $px_1 \in \mathfrak{A}(R)$, and, further, an arbitrary element $a \in R$ can be written in the form $a = Ax_i + y$ for some integer A, some natural number i, and some $y \in \mathfrak{A}(R)$. We shall prove that R is a null ring, by showing $x_k \in \mathfrak{A}(R)$, $k = 1, 2, \ldots$. First note that $p^k x_k \in \mathfrak{A}(R)$ and $x_i - p^k x_{k+i} \in \mathfrak{A}(R)$, $i = 1, 2, \ldots$. Thus

$$x_k a = x_k(Ax_i + y) = Ax_k(x_i - (x_i - p^k x_{k+i})) = Ax_k(p^k x_{k+i}) = 0.$$

Dually $ax_k = 0$, so $x_k \in \mathfrak{A}(R)$. The result follows.

3.5.7 *A total nilpotent algebra of degree at least 3 is not capable.*

Proof Let P be the free algebra on indeterminates
$$\{x_{ij} \mid n \geq i > j \geq 1\}.$$
Then P/K is isomorphic to a total nilpotent algebra, where K is the ideal generated by
$$x_{ij}x_{kl} - \delta_{jk}x_{il}$$
for $n \geq i > j \geq 1$, $n \geq k > l \geq 1$. $\delta_{jk} = 0$ if $j \neq k$, $\delta_{jj} = 1$. Let $N_0 = PK + KP$. Since $x_{n1} \notin K$, it is sufficient to show
$$x_{n1}P + Px_{n1} \subseteq N_0.$$

Claim 1 $x_{n1}x_{ij} \in N_0$, $n \geq i > j \geq 1$.
$x_{ni}x_{i1} - x_{n1} \in K$ so $x_{ni}x_{i1}x_{ij} - x_{n1}x_{ij} \in N_0$. But $x_{i1}x_{ij} \in K$ so $x_{ni}x_{i1}x_{ij} \in N_0$. Thus $x_{n1}x_{ij} \in N_0$.

Claim 2 $x_{ij}x_{n1} \in N_0$, $n \geq i > j \geq 1$.
$x_{n2}x_{21} - x_{n1} \in K$ so $x_{ij}x_{n2}x_{21} - x_{ij}x_{n1} \in N_0$. But $x_{ij}x_{n2} \in K$ so $x_{ij}x_{n2}x_{21} \in N_0$. Thus $x_{ij}x_{n1} \in N_0$.

3.5.8 *Every cyclic nilpotent p-ring S is capable. Let*
$$S = \langle a \mid a^2 = p^m a, \text{ char } a = p^n, 1 \leq m \leq n \rangle.$$
There is only one family of rings R for which $R/\mathfrak{A}(R) \cong S$, represented by the cyclic ring $R = \langle b \mid b^2 = p^m b, \text{ char } b = p^{m+n} \rangle$.

Proof Let P be the free ring with one generator x, and
$$K = \langle p^n x, x^2 - p^m x, x^3 - p^{2m}x, x^4 - p^{3m}x, \ldots \rangle,$$
$$N_1 = P^2 \cap K = \langle p^{n-m}x^2, x^3 - p^m x^2, x^4 - p^{2m}x^2, \ldots \rangle,$$
$$N_0 = xK = \langle p^n x^2, x^3 - p^m x^2, x^4 - p^{2m}x^2, \ldots \rangle.$$
Then $S \cong P/K$, and $K/N_0 = \mathfrak{A}(P/N_0)$, so, by **3.5.1** S is capable. Further, N_1/N_0 is cyclic of order p^m, and hence the ideals N satisfying (1) of **3.5.2** are of the form $N = \langle p^k x^2, N_0 \rangle$, $n - m \leq k \leq n$. Clearly,

then, $p^k x + N \in \mathfrak{A}(P/N)$, so if N is capable then $N = N_0$. Thus only the maximal family exists, represented by P/N_0. Finally let
$$N = \langle x^2 - p^m x \rangle + N_0.$$
Then $N \cap P^2 = N_0$, so by **3.2.3** $P/N \overset{F}{\leftrightarrow} P/N_0$. But
$$P/N \cong R = \langle b \mid b^2 = p^m b, \text{ char } b = p^{m+n} \rangle.$$

3.5.9 *Calculation of the families of rings R for which $R/\mathfrak{A}(R)$ is the null ring $Z11$ of order 4 with an elementary abelian additive group.*

By **3.5.4** $Z11$ is capable. By **3.2.8** each family of rings R with $R/\mathfrak{A}(R) \cong Z11$ contains a 2-algebra, and hence without loss of generality we may perform all computations over $GF(2)$. Let P be the free 2-algebra on indeterminates $\{x, y\}$, $K = N_1 = P^2$, so $N_0 = P^3$. Thus N_1/N_0 has a basis consisting of all monomials of length 2 from P. The automorphism group of P/K is clearly the non-abelian group of order 6, Σ_3. Σ_3 is generated by two involutions whose action on P/K and induced action on N_1/N_0 can be tabulated as follows:

Image of		Induced image of			
x	y	x^2	xy	yx	y^2
$x + y$	y	$x^2 + xy + yx + y^2$	$xy + y^2$	$yx + y^2$	y^2
y	x	y^2	yx	xy	x^2

We now calculate the orbits of subgroups N, $N_0 \le N \le N_1$. The orbits are numbered, and different subgroups N in the same orbit are separated by semicolons. A subgroup N is specified by listing a set of elements $\{w_1, w_2, \ldots, w_k\}$ which, together with N_0, spans N.

Case I $\quad |N/N_0| = 1$
1 $N = N_0$.

Case II $\quad |N/N_0| = 2$
1 $\{x^2\}$; $\{y^2\}$; $\{x^2 + xy + yx + y^2\}$
2 $\{xy\}$; $\{yx\}$; $\{x^2 + xy\}$; $\{x^2 + yx\}$; $\{xy + y^2\}$; $\{yx + y^2\}$
3 $\{xy + yx\}$
4 $\{x^2 + y^2\}$; $\{x^2 + xy + yx\}$; $\{xy + yx + y^2\}$
5 $\{x^2 + xy + y^2\}$; $\{x^2 + yx + y^2\}$.

Case III $\quad |N/N_0| = 4$
1 $\{x^2, xy, x^2 + xy\}$; $\{yx, y^2, yx + y^2\}$;
   $\{x^2 + yx, xy + y^2, x^2 + xy + yx + y^2\}$
2 $\{x^2, yx, x^2 + yx\}$; $\{xy, y^2, xy + y^2\}$;
   $\{x^2 + xy, yx + y^2, x^2 + xy + yx + y^2\}$

3 $\{x^2, y^2, x^2 + y^2\}$; $\{x^2, xy + yx + y^2, x^2 + xy + yx + y^2\}$;
   $\{y^2, x^2 + xy + yx, x^2 + xy + yx + y^2\}$
4 $\{x^2, xy + yx, x^2 + xy + yx\}$; $\{y^2, xy + yx, xy + yx + y^2\}$;
   $\{x^2 + y^2, xy + yx, x^2 + xy + yx + y^2\}$
5 $\{x^2, xy + y^2, x^2 + xy + y^2\}$; $\{y^2, x^2 + xy, x^2 + xy + y^2\}$;
   $\{yx, x^2 + xy + y^2, x^2 + xy + yx + y^2\}$;
   $\{xy, x^2 + yx + y^2, x^2 + xy + yx + y^2\}$;
   $\{y^2, x^2 + yx, x^2 + yx + y^2\}$;
   $\{x^2, yx + y^2, x^2 + yx + y^2\}$
6 $\{xy, yx, xy + yx\}$; $\{xy + yx, xy + y^2, yx + y^2\}$;
   $\{x^2 + xy, x^2 + yx, xy + yx\}$
7 $\{xy, x^2 + yx, x^2 + xy + yx\}$; $\{xy, yx + y^2, xy + yx + y^2\}$;
   $\{x^2 + yx, yx + y^2, x^2 + y^2\}$; $\{x^2 + xy, xy + y^2, x^2 + y^2\}$;
   $\{yx, xy + y^2, xy + yx + y^2\}$; $\{yx, x^2 + xy, x^2 + xy + yx\}$
8 $\{xy, x^2 + y^2, x^2 + xy + y^2\}$;
   $\{yx + y^2, x^2 + xy + yx, x^2 + xy + y^2\}$;
   $\{x^2 + yx, x^2 + xy + y^2, xy + yx + y^2\}$;
   $\{xy + y^2, x^2 + xy + yx, x^2 + yx + y^2\}$;
   $\{yx, x^2 + y^2, x^2 + yx + y^2\}$;
   $\{x^2 + xy, x^2 + yx + y^2, xy + yx + y^2\}$
9 $\{x^2 + y^2, x^2 + xy + yx, xy + yx + y^2\}$
10 $\{xy + yx, x^2 + xy + y^2, x^2 + yx + y^2\}$.

Case IV $|N/N_0| = 8$
   1 $\{x^2, y^2, xy + yx\}$
   2 $\{x^2, xy + yx, yx + y^2\}$; $\{y^2, xy + yx, x^2 + xy\}$;
      $\{yx, xy + yx, x^2 + y^2\}$
   3 $\{x^2, y^2, xy\}$; $\{y^2, x^2 + xy, yx\}$; $\{x^2, yx, xy + y^2\}$;
      $\{y^2, xy, x^2 + yx\}$; $\{x^2, y^2, yx\}$; $\{x^2, xy, yx + y^2\}$
   4 $\{yx, x^2 + y^2, x^2 + xy\}$; $\{xy, x^2 + y^2, x^2 + yx\}$
   5 $\{xy, yx, x^2\}$; $\{xy, yx, y^2\}$; $\{x^2 + yx, x^2 + xy, xy + y^2\}$.

Case V $|N/N_0| = 16$
   1 $N = N_1$.

One can easily check that each ideal N satisfies $\mathfrak{A}(P/N) = K/N$, and hence corresponds to a family, except when $N = N_1$ or when N is in orbit 5, Case IV. These last two orbits do not represent families.

Finally let us note, for each capable N, that $\mathfrak{A}(P/N) = K/N = P^2/N$, and so P/N is a stem 2-algebra in its family, which is unique up to isomorphism. In Case IV, for example, orbits 1, 2, 3, 4 correspond respectively to the 2-algebras $B111$, $C_1 111$, $D_0 111$, and $D_1 111$.

6 Conditions for capability

The object of this section is to find conditions which determine the capability of a ring in terms of the capability of its subrings.

3.6.1 THEOREM *A subdirect sum of capable rings is capable.*

Proof Let S be a subdirect sum of S_1 and S_2. Suppose $S_i \cong R_i/A_i$, where $A_i = \mathfrak{A}(R_i)$, $i = 1, 2$. Since
$$S \subseteq S_1 \oplus S_2 \cong (R_1 \oplus R_2)/(A_1 \oplus A_2)$$
and S is a subdirect sum, there is a subring $R' \supseteq A_1 \oplus A_2$ of $R_1 \oplus R_2$ with $R'/(A_1 \oplus A_2) \cong S$, and homomorphisms φ_i of R' onto R_i, $i = 1, 2$. Clearly $A_1 \oplus A_2 = \mathfrak{A}(R_1 \oplus R_2)$, and since φ_1, φ_2 are onto,
$$A_1 \oplus A_2 = \mathfrak{A}(R').$$
Thus $S \cong R'/\mathfrak{A}(R')$ is capable.

There are many examples to show that capability of a subdirect sum does not imply capability of the summands. A simple one is the p-algebra S with a basis $\{a, b, ab, ba\}$ with $a^2 = b^2 = aba = bab = 0$. To show that S is capable define a p-algebra R with basis $\{x, y, xy, yx, xyx, yxy\}$ with $x^2 = y^2 = (xy)^2 = (yx)^2 = 0$. Clearly R is associative and $S \cong R/\mathfrak{A}(R)$. Moreover, S is isomorphic to a subdirect sum of $S/\langle ab \rangle$ and $S/\langle ba \rangle$. We shall show in **3.6.5** that these summands are not capable. We now prove, however, that in the case of a direct sum the summands are capable. We need the following lemma. Example **3.5.6** shows that some type of finiteness assumption is required in the lemma.

3.6.2 LEMMA *Let $\varphi : P \to R$ be a homomorphism of a free ring P onto a finite ring R, with kernel K. If R is not capable, then there exists $x \in P^2$, $x \notin K$, such that $xP + Px \subseteq N_0 = KP + PK$.*

Proof If $R = 0$, the lemma is vacuous. Suppose R is not capable. To show the existence of $x \in P^2$, $x \notin K$, with $xP + Px \subseteq N_0$, we proceed by induction on $|R|$. Suppose $x \notin K$, $xP + Px \subseteq N_0$, and $\langle \varphi(x) \rangle \cap R^2 = 0$. Then by **1.4.1** R is isomorphic to a subdirect sum of $R/\langle \varphi(x) \rangle$ and the null ring R/R^2. By **3.5.4** and **3.6.1** $R/\langle \varphi(x) \rangle$ cannot be capable. By induction hypothesis there exists $y \in P^2$, $y \notin K + \langle x \rangle$ such that $yP + Py \subseteq (K + \langle x \rangle)P + P(K + \langle x \rangle) = N_0 + (xP + Px)$. By hypothesis $xP + Px \subseteq N_0$. Hence $yP + Py \subseteq N_0$ and $y \in P^2$, $y \notin K$.

3.6.3 THEOREM *A direct sum of finite rings is capable if and only if each summand is capable.*

Proof Let $R = S_1 \oplus S_2$ and suppose S_1 is not capable. We prove R is not capable. Let P_i be the free polynomial ring on a set of indeterminates X_i, with $S_i \cong P_i/K_i$, $i = 1, 2$. Let P be the free polynomial ring on $X_1 \cup X_2$. Then $R \cong P/K$, where
$$K = \langle K_1, K_2, K_1 P_2 + P_2 K_1, K_2 P_1 + P_1 K_2 \rangle.$$
By **3.6.2** there is $x \in P_1^2$, $x \notin K_1$ such that
$$xP_1 + P_1 x \subseteq K_1 P_1 + P_1 K_1 \subseteq N_0 = PK + KP.$$
Let $x = \sum x_j y_j$, $x_j, y_j \in P_1$. Since $y_j P_2 \subseteq K$, and $P_2 x_j \subseteq K$,
$$xP_2 + P_2 x \subseteq N_0.$$
Hence $xP + Px \subseteq N_0$, so R is not capable.

3.6.4 COROLLARY *Capability is a family invariant for finite dimensional algebras.*

Proof This follows from **3.2.7** and the finite dimensional algebra analogues of **3.6.2** and **3.6.3**.

An examination of the nilpotent p-rings of small orders shows that capability severely restricts the structure of the ring. As an example of the sort of restrictions that can be obtained, we shall characterize the capable finite nilpotent p-rings R for which $| R^2 | = p$.

3.6.5 THEOREM *Let R be a finite nilpotent p-ring for which $| R^2 | = p$. Then R is capable if and only if R is isomorphic to the direct sum of a null ring and one of the following rings S:*
(1) $S = \langle \varphi \mid \text{char } \varphi = p^k, \varphi^2 = p^{k-1} \varphi \rangle$, *some integer $k > 0$.*
(2) $S = \langle \varphi, \psi \mid \text{char } \varphi = p^s, \text{char } \psi = p^t, t \leq s, \varphi^2 = p^{t-1} \psi,$
$\varphi\psi = \psi\varphi = \psi^2 = 0 \rangle$.

Proof It is easy to check that the rings S described in (1) and (2) are capable. So, for N null, $N \oplus S$ is capable by **3.6.1**. We now analyze the structure of a capable finite nilpotent p-ring R with $| R^2 | = p$. Choose $\zeta \in R^2$ with char $\zeta = p$. Denote generic elements of R by α, β. Fix a map $P \to R$ of a free ring P onto R, with kernel K, and let $a \to \alpha$, $b \to \beta$, $z \to \zeta$. Let $N_0 = PK + KP$. We require two lemmas.

3.6.6 *Suppose $\alpha^2 = 0$ and $\alpha\beta \neq 0$. Then $z\langle a, b \rangle + \langle a, b \rangle z \subseteq N_0$.*

Proof Since $\alpha\beta \neq 0$ and $| R^2 | = p$, there are integers A, B, C such that $\beta\alpha = A \alpha\beta$, $\beta^2 = B \alpha\beta$, and $\zeta = C \alpha\beta$. $\alpha^2 = 0$ means $a^2 \in K$, so $a^2 (Cb) = a(Cab) \in N_0$, so $az \in N_0$. Since $ba - Aab \in K$ and $a(ab) \in N_0$, $a(ba) = (ab)a \in N_0$. Hence $za \in N_0$. Since $b^2 - Bab \in K$ and $a(ab) \in N_0$, $ab^2 = (ab)b \in N_0$. Hence $zb \in N_0$. Finally
$$(ba - Aab) \in N_0,$$
and $ab^2 \in N_0$, so $(ba)b = b(ab) \in N_0$. Hence $bz \in N_0$.

3.6.7 *Suppose* $\alpha^2 \neq 0$, $\beta^2 \neq 0$, *and* $\alpha\beta = 0$. *Then*
$$z\langle a, b \rangle + \langle a, b \rangle z \subseteq N_0.$$

Proof There are integers $A \not\equiv 0 \pmod{p}$ and B such that $\beta^2 = A\alpha^2 \neq 0$ and $\beta\alpha = B\alpha^2$. From $ab \in K$ and
$$a(b^2 - Aa^2) = (ab)b - Aa^3 \in N_0$$
follows $az \in N_0$, $za \in N_0$. Since $a^2b = a(ab) \in N_0$, $zb \in N_0$. Finally, $(ba - Ba^2)a \in N_0$ and $az \in N_0$ imply $(ba)a = ba^2 \in N_0$, so $bz \in N_0$.

Proof of **3.6.5**, *continued* We shall construct a set of elements $\{\alpha, \beta_1, \ldots, \beta_n\}$ which is a basis for R, and for which $\beta_i \in \mathfrak{A}(R)$, $1 \leq i \leq n$. Since R is capable and not null, there is an element $\alpha \in R$, $\alpha \notin \mathfrak{A}(R)$, for which $za \notin N_0$ or $az \notin N_0$. Since $\alpha \notin \mathfrak{A}(R)$, there exists β such that $\alpha\beta \neq 0$ or $\beta\alpha \neq 0$. Then **3.6.6** or its dual implies $\alpha^2 \neq 0$. Without loss of generality $\alpha^2 = \zeta$. Select β_i, $1 \leq i \leq m$, so that $\{\alpha, \beta_1, \ldots, \beta_m\}$ spans R. Let $\alpha\beta_i = A_i\zeta$. By replacing each β_i by $\beta_i - A_i\alpha$, we can assume that $\alpha\beta_i = 0$, $1 \leq i \leq m$. Then **3.6.7** implies $\beta_i^2 = 0$. Then **3.6.6** implies $\beta_i\alpha = 0$. Suppose $\beta_i\beta_j = A\zeta$. Then $a(b_ib_j - Az) \in N_0$, and $ab_i \in K$, so $Aaz \in N_0$. Dually $Aza \in N_0$. But one of $az \notin N_0$ and $za \notin N_0$ holds by the choice of α, so $A \equiv 0 \pmod{p}$. Thus $\beta_i \in \mathfrak{A}(R)$, $1 \leq i \leq m$, and so $| R/\mathfrak{A}(R) | = p$.

Now choose a basis $\{\gamma_1, \ldots, \gamma_n\}$ for R. Order the basis so that $\gamma_1 \notin \mathfrak{A}(R)$ and has minimal characteristic among the $\gamma_i \notin \mathfrak{A}(R)$. Since $| R/\mathfrak{A}(R) | = p$, there are integers A_j, $0 \leq A_j < p$, $2 \leq j \leq n$, such that $\gamma_j \equiv A_j\gamma_1 \pmod{\mathfrak{A}(R)}$. Replace γ_j by $\gamma_j - A_j\gamma_1$, which makes $\gamma_j \in \mathfrak{A}(R)$, $2 \leq j \leq n$. Without loss of generality, $\gamma_1^2 = \zeta$. Next, since char $\zeta = p$, we can alter the basis so that ζ is a natural multiple of basis element γ_i for one of $i = 1$ or $i = 2$. $i = 1$ leads directly to case (1) of the conclusion. Suppose $i = 2$. Let char $\gamma_1 = p^s$, char $\gamma_2 = p^t$. To show that case (2) holds we must show $t \leq s$. Let $c_i \to \gamma_i$, $1 \leq i \leq n$, under the map of the free ring P onto R. Suppose $t > s$. Then $p^{t-1}c_1 \in K$ so $(p^{t-1}c_1)c_2 = c_1(p^{t-1}c_2) \in N_0$, so $c_1z \in N_0$. Similarly $zc_1 \in N_0$. This contradicts the capability of R, since $\gamma_i \in \mathfrak{A}(R)$ and $\zeta \in R^2$ implies $c_iz, zc_i \in N_0$, $2 \leq i \leq n$. Thus $t \leq s$, and case (2) of the conclusion holds.

CHAPTER IV

The subring structure of nilpotent rings

THIS CHAPTER continues the study of nilpotent ring analogues of finite p-group results. While the last chapter discussed some "algebraic" properties of nilpotent rings, the present chapter is concerned with some "arithmetical" properties. We shall examine several numerical invariants connected with a nilpotent p-ring and with its automorphism group. In §1 the generators of finite rings are studied. The Frattini subring of a ring is introduced, and a ring analogue of the Burnside basis theorem is proved. In §2 the automorphism group of a finite nilpotent p-ring is examined. Anzahl results are given in §3. It is shown that the number of subrings, right ideals, and two-sided ideals of a given order in a finite nilpotent p-ring is congruent to 1 mod p. In §§4 and 5 a characterization is given of the class of nilpotent p-rings with a unique subring of a given order. In §6, finally, the p-rings in which every subring is an ideal are studied.

1 On generating rings

The (right) Frattini subring Φ_R of a ring R is defined to be the intersection of all maximal right ideals of R, provided such exist. If not, $\Phi_R = R$. In this section we shall denote the smallest right ideal containing a subset S of a ring by $[S]_r$. To find the relation of the Frattini subring to the generation of R, we shall call an element $x \in R$ a *right non-generator* of R if, whenever S is a subset of R with $R = [S, x]_r$, then $R = [S]_r$. Note that x is a nongenerator only if x may be omitted from *every* such set.

4.1.1 THEOREM *The right Frattini subring Φ_R of a ring R is exactly the set of right non-generators of R.*

Proof Choose $x \in R$. If there is a maximal right ideal S of R with $x \notin S$, then $[S, x]_r$ properly contains S, so $[S, x]_r = R$ by the maximality of S. Hence a nongenerator of R must be contained in every maximal right ideal S, hence in Φ_R.

We must now show that $x \in \Phi_R$ implies x is a nongenerator. Suppose $x \in R$ and S is a subset of R with $[S, x]_r = R$, $[S]_r = T \neq R$. Certainly $x \notin T$. Then by Zorn's Lemma there exists a right ideal M maximal with respect to $M \supseteq T$, $x \notin M$. Any right ideal of R which properly contains M must contain x, and $[M, x]_r \supseteq [S, x]_r$, so M is a maximal right ideal of R which does not contain x. Hence $x \notin \Phi_R$.

Remark The proofs of **4.1.1** and **4.1.3** contain the only uses of Zorn's Lemma or other form of the axiom of choice to be found in this book.

4.1.2 THEOREM *Let N be the Jacobson radical of a ring R. Then $RN \subseteq \Phi_R \subseteq N$. Suppose R satisfies the descending chain condition for one-sided ideals. Then Φ_R is nilpotent, and Φ_R contains R^2 if and only if R is nilpotent.*

Proof To see that Φ_R is contained in N observe that Φ_R is the intersection of all maximal right ideals of R, while N is the intersection of all maximal right ideals which are "modular" (see Jacobson [3], p. 9).

$RN \subseteq \Phi_R$ is trivial if $\Phi_R = R$, so consider the case when Φ_R is proper in R. Suppose $x \in R$ has the property that there exists $y \in R$ with $yx \notin \Phi_R$. Then there exists a maximal right ideal T of R with $yx \notin T$. T defines a simple right R-module $M \simeq R^+/T^+$, and under the natural R-homomorphism φ of R onto M, $\varphi(y) \neq 0$ and $\varphi(yx) \neq 0$. Thus $MR = M$. Thus an element $z \in R$ exists such that $\varphi(yxz) = -\varphi(y)$. Suppose that xz is right quasi-regular, so that there exists $b \in R$ such that $xz + b + xzb = 0$. Then

$$\varphi(yxz) = -\varphi(yb) - \varphi(yxzb)$$
$$= -\varphi(yb) + \varphi(yb) = 0,$$

a contradiction. Thus xz is not right quasi-regular, so $x \notin N$. Hence for every element $x \in N$ and $y \in R$, $yx \in \Phi_R$. Thus $RN \subseteq \Phi_R$.

Now suppose R satisfies the descending chain condition. Then N is nilpotent, and $\Phi_R \subseteq N$, so Φ_R is nilpotent. If R is nilpotent then by **1.3.5** $R^2 \subseteq \Phi_R$. If, on the other hand, $R^2 \subseteq \Phi_R$, then R/Φ_R, along with R/N, is null. But R/N is semi-simple, so $R = N$ is nilpotent.

The following simple observation will frequently be useful.

4.1.3 *If R is a nilpotent p-ring then $\Phi_R = R^2 + pR$.*

Proof By **1.3.5** $R^2 \subseteq \Phi_R$ and by **1.3.6** $pR \subseteq \Phi_R$. Let $T = R/(R^2 + pR)$. Then T is a null p-algebra, so clearly the intersection of the maximal right ideals of T is 0, which means that $\Phi_R \subseteq R^2 + pR$.

If the Frattini subalgebra of an algebra A is defined as the intersection of the maximal right ideals of A, if any, then the obvious analogues of Theorems **4.1.1** and **4.1.2** hold, with finite dimensionality in place of descending chain condition. Note that the Frattini subalgebra of a nilpotent algebra A is just A^2.

Historical remark The right Frattini subring was first introduced by L. Fuchs [1], and, for rings with an identity, characterized as the Jacobson radical. An extension of Fuchs' result has been obtained by A. Kertész [1]. A two-sided ideal version of the Frattini subring has been introduced by H. Bechtell [1], and its relations with the one-sided version and with the radical of the ring examined. Theorem **4.1.2** is due to N. Jacobson [2]. For Lie algebras a Frattini subalgebra has been studied by E. Marshall [1].

In the sequel we shall frequently use the observation that, in a nilpotent ring, every maximal two-sided (or one-sided) ideal is already a maximal subring (see **1.3.7**), so that the Frattini subring is the intersection of all maximal subrings.

4.1.4 Burnside Basis Theorem *Let R be a finite nilpotent p-ring. Then $A = R/\Phi_R$ is a null ring, and A^+ is elementary abelian. Let $[R:\Phi_R] = p^d$. Then any set of elements of R which generates R contains a subset of d elements, $\{r_1, \ldots, r_d\}$, which generates R. In the canonical homomorphism of R onto A the elements $r_1, \ldots, r_d$ map onto a basis of A. If, conversely, $r_1 + \Phi_R, \ldots, r_d + \Phi_R$ form a basis of A, then $\{r_1, \ldots, r_d\}$ generates R.*

Proof By **1.3.5** and **1.3.6** A is a null ring and A^+ is elementary abelian. Thus the images under the canonical homomorphism $R \to A$ of any generating set for R must contain a basis for A. Let $\{r_1, \ldots, r_d\}$ be a set of elements whose images form a basis for A. Suppose $r_1, \ldots, r_d$ generate a proper subring of R. By **1.3.7** this subring is contained in a maximal ideal I, which contains Φ_R. Thus $r_1 + \Phi_R, \ldots, r_d + \Phi_R$ are in I/Φ_R, which is proper in A. This contradicts the assumption that the images of $r_1, \ldots, r_d$ form the basis of A. This completes the proof.

4.1.5 Corollary *A finite nilpotent ring R is a power ring if and only if R contains a unique maximal subring.*

2 Automorphisms

This section studies the group of automorphisms of a finite nilpotent p-ring R by the use of the Frattini subring Φ_R and the Burnside basis theorem. First are studied the automorphisms of R which fix R/Φ_R elementwise, and a bound is obtained on the order of the automorphism group of R. These results are analogous to the p-group results of P. Hall [1]. Next the "inner automorphisms" of R are discussed, and the concepts illustrated by their application to total nilpotent algebras. Finally, a bound on the class of the p-group of automorphisms which fix R/Φ_R elementwise is obtained.

We begin with a simple observation which will be of frequent use in this section.

4.2.1 LEMMA *Let R be a ring and α an automorphism of R. If $\{r_1, \ldots, r_d\}$ is a generating set for R then α is completely determined by the values of $r_i{}^\alpha$, $1 \leq i \leq d$. If I is an ideal of R then α fixes R/I elementwise if and only if $r_i{}^\alpha - r_i \in I$, $1 \leq i \leq d$.*

4.2.2 THEOREM *Let R be a ring and $\mathcal{M}$ the set of automorphisms of R which fix R/Φ_R elementwise. Then $\mathcal{M}$ is a normal subgroup of the automorphism group of R.*

Proof $\mathcal{M}$ is clearly a subgroup of the automorphism group $\mathcal{A}$ of R. Since Φ_R is the intersection of the maximal right ideals of R, $x \in \Phi_R$ implies $x^\alpha \in \Phi_R$, all $\alpha \in \mathcal{A}$. Choose $\mu \in \mathcal{M}$, $\alpha \in \mathcal{A}$, $r \in R$. Then

$$\begin{aligned}
r^{\alpha\mu\alpha^{-1}} &= (r + s)^{\mu\alpha^{-1}} && \text{for some } s \in R \\
&= (r + s + x)^{\alpha^{-1}} && \text{for some } x \in \Phi_R \\
&= r + y && \text{for } y = x^{\alpha^{-1}} \in \Phi_R.
\end{aligned}$$

Thus $\alpha\mu\alpha^{-1} \in \mathcal{M}$, and $\mathcal{M}$ is normal in $\mathcal{A}$.

4.2.3 THEOREM *Let R be a nilpotent ring of order p^n, and let $[R:\Phi_R] = p^d$. Then the order of the automorphism group $\mathcal{A}$ of R divides $p^{d(n-d)} \theta(p^d)$, where*

$$\theta(p^d) = (p^d - 1)(p^d - p) \ldots (p^d - p^{d-1}).$$

The order of the normal subgroup $\mathcal{M}$ of automorphisms of R fixing R/Φ_R elementwise divides $p^{d(n-d)}$.

Proof We first count the ways of choosing ordered generating sets $X = (a_1, \ldots, a_d)$ for R. Such a set determines a basis

$$a_1 + \Phi_R, \ldots, a_d + \Phi_R$$

for R/Φ_R. These ordered bases may be chosen in $\theta(p^d)$ ways. Each element of such a basis corresponds to $|\Phi_R| = p^{n-d}$ elements of R.

Hence there are $p^{d(n-d)}\theta(p^d)$ sets X. Every automorphism $\alpha \in \mathscr{A}$ induces a permutation of the sets X, and if $\alpha \neq 1$ then α fixes no set X. Thus in the representation of $\mathscr{A}$ as permutations of the sets X, each orbit has length $|\mathscr{A}|$, so $|\mathscr{A}|$ divides $p^{d(n-d)}\theta(p^d)$. In the same way $\mathscr{M}$ permutes regularly the $p^{d(n-d)}$ sets X which map onto a fixed basis $a_1 + \Phi_R, \ldots, a_d + \Phi_R$ under $R \to R/\Phi_R$, and hence $|\mathscr{M}|$ divides $p^{d(n-d)}$. This completes the proof.

If ring R contains an identity and $a \in R$ is invertible then the map $r \to a^{-1}ra$, all $r \in R$, is an automorphism of R. Thus, if x is a quasi-regular element of an arbitrary ring R, and if $x + \bar{x} + x\bar{x} = 0$, then the map $r \to r + \bar{x}r + rx + \bar{x}rx = (1 + \bar{x})r(1 + x)$, all $r \in R$, is an automorphism of R. An automorphism obtained from a quasi-regular element in this way is called an *inner automorphism*. This concept is due to A. I. Mal'cev [1]. The next result follows immediately.

4.2.4 THEOREM *The inner automorphisms of a ring R form a normal subgroup of the automorphism group of R, and fix R/R^2 elementwise.*

4.2.5 COROLLARY *If R is a nilpotent ring then the inner automorphisms of R fix R/Φ_R elementwise.*

Remark Whereas every finite p-group must possess an outer automorphism, there are finite nilpotent p-rings for which all automorphisms are inner. An example is the 2-algebra D_0111, which has a basis $\{x, y, z\}$ such that $xy = z$ and all other products of basis members are 0.

To illustrate the concepts introduced we now study the automorphisms of total nilpotent algebras (see §2.1). Recall that the interest in total nilpotent algebras arises from the fact that every finite dimensional nilpotent algebra occurs as a subalgebra of a total nilpotent algebra. The results to follow are due to R. Dubisch and S. Perlis [1]. Let T be the total nilpotent algebra of dimension n over a field F, with a basis of matric units e_{ij}, $n \geq i > j \geq 1$, where matrix e_{ij} has a 1 in position ij and 0's elsewhere.

Let $\mathscr{A}$ be the automorphism group of T, $\mathscr{M}$ the normal subgroup fixing $T/\Phi_T = T/T^2$ elementwise, and $\mathscr{I}$ the normal subgroup of inner automorphisms. Dubisch and Perlis call $\mathscr{M}$ the group of "monic" automorphisms of T.

Let $d = \sum \varphi_i e_{ii}$ be a non-singular $n \times n$ diagonal matrix over F. The map $x^\delta = d x d^{-1}$, all $x \in T$, is an automorphism of T. The group of all such δ is called the "diagonal" automorphism group $\mathscr{D}$ of T.

4.2.6 THEOREM *The automorphism group $\mathscr{A}$ of T is a semi-direct product of the normal subgroup $\mathscr{M}$ of monic automorphisms by the subgroup $\mathscr{D}$ of diagonal automorphisms.*

Proof The terms of the left and right annihilator series $L_1 \subset L_2 \subset \ldots L_{n-1}$ and $R_1 \subset R_2 \subset \ldots R_{n-1}$ described in section (2.1) are fixed under all automorphisms of T. Since, for $1 \leq k \leq n - 1$, $L_k \cap R_{n-k} = \langle e_{ij} \mid j < i, i \leq k + 1, j \leq k \rangle$, it follows that for $\alpha \in \mathscr{A}$, $e_{k+1 k}{}^\alpha = \varphi_k e_{k+1 \ k} + x$ for some $\varphi_k \in F$, $x \in T^2$. From $e_{k+1 \ k} \notin T^2$ follows $\varphi_k \neq 0$. Let $\varphi_n = 1$ and define the diagonal matrix $d = \sum_i \varphi_i e_{ii}$. Define δ by $x^\delta = d \, x \, d^{-1}$, $x \in T$. Then $\delta^{-1}\alpha \in \mathscr{M}$. Surely $\mathscr{D} \cap \mathscr{M} = 1$, and so $\mathscr{A}$ is a semi-direct product of $\mathscr{M}$ by $\mathscr{D}$.

Remark A decomposition similar to that of **4.2.6** has been found for free nilpotent algebras by M. Okuzumi [1].

4.2.7 THEOREM *The group $\mathscr{M}$ of monic automorphisms of T is isomorphic to the direct product of $\mathscr{I}$, the group of inner automorphisms of T, and $\mathscr{N}$, the group of automorphisms of T which fix $\mathfrak{A}_l(T) + \mathfrak{A}_r(T)$ elementwise.*

4.2.8 THEOREM *The group $\mathscr{N}$ of automorphisms of T which fix $\mathfrak{A}_l(T) + \mathfrak{A}_r(T)$ elementwise consists of the automorphisms ν defined by $e_{21}^\nu = e_{21}$, $e_{n \ n-1}^\nu = e_{n \ n-1}$, $e_{k \ k-1}^\nu = e_{k \ k-1} + \varphi_k e_{n1}$, for some $\varphi_k \in F$, $3 \leq k \leq n - 1$.*

The proofs of **4.2.7** and **4.2.8** involve rather detailed computation, and may be found in Dubisch and Perlis [1]. The above results have been extended by M. Moriya [1] to certain rings of endomorphisms of arbitrary finite abelian p-groups.

We now return to the more general study of automorphisms. If I is an ideal of a ring R, we now define $\text{Aut}(R;I)$ to be the group of all automorphisms of R which leave R/I fixed elementwise. For R a finite nilpotent p-ring we shall obtain a bound of the class of the p-group $\mathscr{P} = \text{Aut}(R;\Phi_R)$. These results are analogous to those obtained by H. Liebeck [1] for p-groups.

4.2.9 THEOREM *Let R be a finite p-ring, nilpotent of exponent e, for which $\Phi_R \neq 0$. Let R^i/R^{i+1} have characteristic p^{m_i}, $i = 1, \ldots, e - 1$. Then the class of $\mathscr{P} = \text{Aut}(R;\Phi_R)$ does not exceed*

$$\lambda(R) = \left(\sum_{i=1}^{e-1} m_i \right) - 1.$$

This theorem will follow by induction from the next result.

4.2.10 THEOREM *Let R be a finite p-ring, nilpotent of exponent e, for which $\Phi_R \neq 0$. Let R^{e-1} have characteristic p^m and let $N = p^{m-1} R^{e-1}$. Then*

(i) the ideal N is elementwise fixed by $\mathscr{P} = \mathrm{Aut}(R;\Phi_R)$.
(ii) $\mathscr{L} = \mathrm{Aut}(R;N)$ is in the center of $\mathscr{P}$.
(iii) $\mathscr{L}$ has order p^{rd}, where p^r is the order of N, and $p^d = [R:\Phi_R]$.
(iv) $\mathscr{P}/\mathscr{L}$ is isomorphic to the subgroup $\mathscr{Q}$ of automorphisms from $\mathrm{Aut}(R/N;\Phi_R/N)$ which can be extended to R.

The proof of **4.2.10** requires two lemmas.

4.2.11 LEMMA *With R as in **4.2.10** if $\alpha \in \mathscr{P}$ and $x \in R^i$ for some i, $1 \leq i \leq e - 1$, then $x^\alpha - x \in pR^i + R^{i+1}$.*

Proof The lemma is true for $i = 1$, since $\Phi_R = pR + R^2$ and $\alpha \subset \mathrm{Aut}(R;\Phi_R)$. Assume the result for $i < j$. Let $x \in R^j$. Express x as a sum of products

$$x = \sum_r y_r z_r$$

where the $y_r \in R^{j-1}$, $z_r \in R$. Then $x^\alpha = \sum y_r^\alpha z_r^\alpha = \sum (y_r + u_r)(z_r + v_r)$, where, by induction hypothesis, $u_r \in pR^{j-1} + R^j$ and $v_r \in pR + R^2$. Thus $x^\alpha - x = \sum (u_r z_r + y_r v_r + u_r v_r) \in pR^j + R^{j+1}$, q.e.d.

4.2.12 LEMMA *With the notation of **4.2.10**, every automorphism $\zeta \in \mathscr{L}$ leaves Φ_R elementwise fixed.*

Proof For any $x \in R$, $x^\zeta - x \in N$. Thus from $pN = 0$ follows $0 = p(x^\zeta - x) = (px)^\zeta - px$, so $\mathscr{L}$ fixes pR elementwise. Similarly, since $NR = RN = 0$, $\mathscr{L}$ fixes R^2, hence $\Phi_R = R^2 + pR$, elementwise.

*Proof of **4.2.10***
(i) Suppose $x \in N$ and $\alpha \in \mathscr{P}$. Then $x = p^{m-1} y$ for some $y \in R^{e-1}$, and so, by **4.2.11**, $x^\alpha - x = p^{m-1}(y^\alpha - y) \in p^{m-1}(pR^{e-1} + R^e) = 0$.
(ii) Let $x \in R$, $\alpha \in \mathscr{P}$, and $\zeta \in \mathscr{L}$. Then

$$\begin{aligned}(x^\alpha)^\zeta &= (x + u)^\zeta \text{ for some } u \in \Phi_R \\ &= x^\zeta + u \text{ by } \textbf{4.2.12} \\ &= x + u + v \text{ for some } v \in N\end{aligned}$$

while $\quad (x^\zeta)^\alpha = (x + v)^\alpha = x^\alpha + v \text{ [by (i)]} = x + v + u$.
Thus $\alpha\zeta = \zeta\alpha$.
(iii) Let $x_1, \ldots, x_d$ generate R, and $y_1, \ldots y_d$ be arbitrary elements of N.
Then the mapping

$$x_i \rightarrow x_i + y_i, \; i = 1, \ldots, d \qquad (*)$$

E

defines an automorphism $\zeta \in \mathscr{L}$. But every automorphism $\zeta \in \mathscr{L}$ induces a mapping of the form (*). Thus $| \mathscr{L} | = p^{rd}$.

(iv) Let $\alpha \in \mathscr{P}$. Defining $(x + N)^\alpha = x^\alpha + N$ for all $x \in R$ gives, by (i), a homomorphism f of $\mathscr{P}$ into $\mathscr{Q}$. If $(x + N)^\alpha = x + N$, all x, then $x^\alpha - x \in N$, so $\alpha \in \mathscr{L}$. Thus the kernel of f is $\mathscr{L}$. Now consider an automorphism $\beta' \in \mathscr{Q}$. Let β be an extension of β' to R, $\beta \in \mathscr{P}$. Then $f\beta \in \mathscr{Q}$, and for $x \in R$, $(x + N)^{f\beta} = x^\beta + N = (x + N)^{\beta'}$ so the homomorphism f is onto $\mathscr{Q}$. This completes the proof of **4.2.10**.

Proof of **4.2.9** $\lambda(R) = 0$ implies $e = 2$ and $m_1 = 1$, hence $\Phi_R = 0$, contrary to assumption. If $\lambda(R) = 1$, then either $e = 2$, $m_1 = 2$, or else $e = 3$, $m_1 = m_2 = 1$. In either case, for $N = p^{me-1-1} R^{e-1}$, $N = \Phi_R$. Hence, by (ii) of **4.2.10**, $\mathscr{P}$ is abelian. We now use induction on $\lambda(R)$. Let $N = p^{me-1-1} R^{e-1}$. By **4.2.10** $\mathscr{L} = \mathrm{Aut}(R;N)$ is central in $\mathscr{P}$ and $\mathscr{P}/\mathscr{L}$ is isomorphic to a subgroup $\mathscr{Q}$ of $\mathrm{Aut}(R/N; \Phi_R/N)$. Since $\lambda(R/N) = \lambda(R) - 1 < \lambda(R)$, by induction hypothesis the class of $\mathscr{Q}$ does not exceed $\lambda(R/N)$. Thus the class of $\mathscr{P}$ does not exceed $\lambda(R) = \lambda(R/N) + 1$.

Remark   The bounds given in Theorems **4.2.9** and **4.2.10** are best possible, at least in the case of finite algebras, as shown both by total nilpotent algebras and by the following example.

4.2.13 EXAMPLE   *The automorphism group* $\mathrm{Aut}(R;R^2)$ *of a free nilpotent algebra* R *over a field has the largest possible class.*

Let w_i, $i = 1, 2, \ldots, n$, be generators of a free nilpotent algebra R of exponent e. We shall require $n \geq 2$, $e \geq 3$. Let $\mathscr{P} = \mathrm{Aut}(R;R^2)$. The bound for the class of $\mathscr{P}$ given in **4.2.9** is $e - 2$. For $e = 3$, $\mathscr{P}$ clearly attains this bound. That $\mathscr{P}$ attains this bound for $e > 3$ follows by a simple induction on e from the following computation.

CLAIM 1   *Let* $Z(\mathscr{P})$ *be the center of* $\mathscr{P}$. *Then* $Z(\mathscr{P}) = \mathscr{L}$, *where* $\mathscr{L} = \{\alpha \in \mathscr{P} \mid w^\alpha - w \in R^{e-1}, \text{all } w \in R\}$.

Proof   Certainly $\mathscr{L} \subseteq Z(\mathscr{P})$. Note that, for $e = 3$, $\mathscr{L} = Z(\mathscr{P})$ is trivial. Thus to show, for $e > 3$, that $Z(\mathscr{P}) \subseteq \mathscr{L}$ we can suppose by induction on e that $Z(\mathscr{P}) \subseteq \mathscr{L}_1 = \{\alpha \in \mathscr{P} \mid w^\alpha - w \in R^{e-2}, \text{all } w \in R\}$. Suppose $\alpha \in Z(\mathscr{P})$. Let $w_i^\alpha - w_i = x_i$, $1 \leq i \leq n$. Since
$$\mathscr{L} \subseteq Z(\mathscr{P}) \subseteq \mathscr{L}_1$$
we may assume that x_i is homogeneous of degree $e - 2$ or else $x_i = 0$, $1 \leq i \leq n$.

Now let us define an automorphism $\beta \in \mathscr{P}$ by setting $w_r^\beta = w_r + w_s w_t$ for some r, $1 \leq r \leq n$, and $w_m^\beta = w_m$, all $m \neq r$, $1 \leq m \leq n$. Then

$w_r{}^{\alpha\beta} = w_r{}^{\beta\alpha}$ gives

$$x_r^{\beta} - x_r = w_s x_t + x_s w_t \tag{1}$$

and $w_m^{\alpha\beta} = w_m^{\beta\alpha}$, $m \neq r$, gives

$$x_m^{\beta} = x_m, \; m \neq r. \tag{2}$$

To prove Claim 1 we next establish

CLAIM 2   *Suppose w_r occurs in a term of x_m expressed as a linear combination of monomials of length $e - 2$, for some $m \neq r$. Then w_r occurs at least twice in that term.*

Proof Consider condition (2) for $s = t = r$. If any term q of x_m involves w_r at least twice, then $q^{\beta} - q$ involves w_r at least three times, by the definition of β, since $q \in R^{e-2}$. The terms of x_m which involve w_r exactly once then lead to an equation of the form $\sum_i X_i \, w_r^2 \, Y_i = 0$. Since the monomials of length $e - 1$ are all linearly independent, the coefficient of each monomial which appears in $\sum X_i \, w_r^2 \, Y_i$ must be 0. Hence the coefficient of each monomial which appears in $\sum X_i \, w_r \, Y_i$ must also be 0, $\sum X_i \, w_r \, Y_i = 0$, and Claim 2 is proved.

CLAIM 3   *w_r occurs in no term of x_m if $m \neq r$.*

Proof Let $s = t = m$. Then condition (2) implies that w_r does not occur in x_m.

Proof of Claim 1, continued By Claim 3, $x_r = A_r \, w_r^{e-2}$ for all r, $1 \leq r \leq n$, some A_r in the field. Then condition (1) implies

$$w_s \, A_t \, w_t^{e-2} + A_s \, w_s^{e-2} w_t = x_r^{\beta} - x_r = A_r (w_r + w_s w_t)^{e-2} - x_r,$$

so

$$w_s \, A_t \, w_t^{e-2} + A_s \, w_s^{e-2} w_t = x_r^{\beta} - x_r = A_r \left((w_r + w_s w_t)^{e-2} - w_r^{e-2}\right). \tag{3}$$

Let $s = t \neq r$. Then the left side does not involve x_r, while the right side does, so $A_r = 0$, all r, $1 \leq r \leq n$, q.e.d.

3 Enumeration results

The results of this section depend upon the following lemma, which is essentially the enumeration principle of Philip Hall ([1], Theorem 1.4).

4.3.1 LEMMA   *Let A be a finite p-group, $\mathscr{M}$ the set of maximal subgroups of A which contain a fixed subgroup $B \neq A$. Let $\mathscr{C}$ be any class whose members are subsets of A, and let each member of $\mathscr{C}$ be contained in at least one member of $\mathscr{M}$. Let $n(M)$ be the number of members of $\mathscr{C}$*

which are contained in M for each $M \in \mathcal{M}$. Then the number of members of $\mathcal{C}$ is congruent to $\sum_{M \in \mathcal{M}} n(M)$ (mod p).

Proof For $C \in \mathcal{C}$ let $M_0 = \cap \{M \in \mathcal{M} \mid C \subseteq M\}$, and let $i(C)$ denote the number of subgroups $M \in \mathcal{M}$ for which $C \subseteq M$. Then $i(C)$ counts the maximal subgroups of the elementary abelian p-group A/M_0, and so $i(C) \equiv 1 \pmod{p}$, all $C \in \mathcal{C}$. But the identity

$$\sum_{C \in \mathcal{C}} i(C) = \sum_{M \in \mathcal{M}} n(M)$$

follows from counting all of the containment relations between $C \in \mathcal{C}$ and $M \in \mathcal{M}$ in the two ways indicated in the summations, so

$$\sum_{C \in \mathcal{C}} 1 \equiv \sum_{C \in \mathcal{C}} i(C) = \sum_{M \in \mathcal{M}} n(M) \pmod{p}.$$

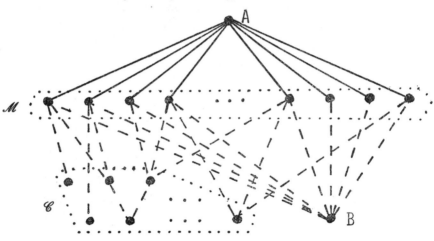

FIGURE 4

4.3.2 THEOREM *Let R be a nilpotent ring of order p^n. Let S be a subring of R, of order p^s. Then, for $s \leq t \leq n$, the number of subrings of R of order p^t which contain S is congruent to 1 (mod p).*

Proof If $S = R$, the result is trivial. Suppose $S \neq R$. We proceed by induction on n. Let $\mathcal{M}$ be the set of maximal subgroups of R^+ which contain $B^+ = (S + R^2)^+$. By **1.3.5** and **1.3.7**, $\mathcal{M}$ is non-empty. Letting $\mathcal{C} = \{S\}$ in **4.3.1**, we see that the number of members of $\mathcal{M}$ is congruent to 1 (mod p). If $t = n$, the result is trivial. Suppose $t < n$. Let $\mathcal{C} = \{T \mid S \subseteq T, \mid T \mid = p^t, T \text{ is a subring of } R\}$. By **1.3.7** and **1.3.5** each $T \in \mathcal{C}$ is contained in some $M \in \mathcal{M}$. Let $n(M)$ be the number of members of $\mathscr{C}$ contained in M, for each $M \in \mathscr{M}$. By induction, $n(M) \equiv 1 \pmod{p}$. Hence, by **4.3.1**, the number of members of $\mathscr{C}$ is congruent to 1 $\pmod{p}$.

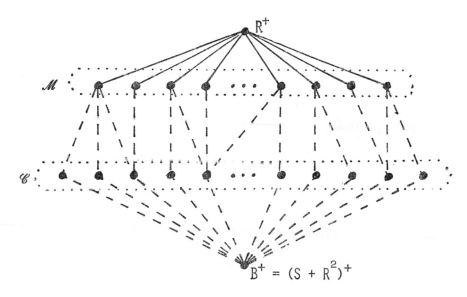

FIGURE 5

It is well known that the number of normal subgroups of a given order in a finite p-group is congruent to 1 $\pmod{p}$. For rings there are several analogous results.

4.3.3 THEOREM   *Let P be a right module of order p^n of a nilpotent ring R. The number of submodules of P of order p^k, $0 \le k \le n$, is congruent to 1 $\pmod{p}$.*

4.3.4 THEOREM   *Let I be a right ideal of order p^m of a nilpotent ring R of order p^n. The number of right ideals of R of order p^k which contain I (which are contained in I), $m \le k \le n$ $(0 \le k \le m)$, is congruent to 1 mod p.*

4.3.5 THEOREM   *Let I be a two-sided ideal of order p^m of a nilpotent ring R of order p^n. The number of two-sided ideals of R of order p^k which contain I (which are contained in I), $m \le k \le n$ $(0 \le k \le m)$, is congruent to 1 mod p.*

The proofs of these results are similar to that of **4.3.2**.

Remark 1   No analogue of the theorem of Kulakoff seems to hold for nilpotent rings. For example, the rings with basis a, b, such that char $a =$ char $b = p^2$, $a^2 = -b^2 = pa$, and $ab = ba = 0$, have $3p + 1$ subrings of order p^2 if $p \neq 2$, and 5 if $p = 2$. A finite non-abelian p-group must contain an abelian subgroup properly larger than its center, but nilpotent p-rings, $A2$ for example, exist which have no null subring larger than the annihilator.

Remark 2   Note that the Anzahl theorems fail to hold for non-nilpotent p-rings. For example, define $R = R_1 \oplus R_2$ where
$$R_1 = \langle a \mid \text{char } a = p,\ a^2 = a \rangle$$
and $R_2 = \langle b \mid \text{char } b = p,\ b^2 = 0 \rangle$. Then R_1 and R_2 are the only two subrings (and ideals) of R of order p, and $2 \not\equiv 1 \pmod{p}$ for any prime p.

4 Nilpotent rings with only one subring of a given order

It is well known that a finite p-group G which contains only one subgroup S of a given order, $1 \neq S \neq G$, must be cyclic, or else $\mid S \mid = 2$ and G is generalized quaternion (Burnside [1], 131–132). In this section and the next we obtain a characterization of the nilpotent rings and algebras satisfying the analogous condition. Although the algebra result follows from the ring result in the usual way, we shall give an independent proof to illustrate the general ideas used while avoiding much detail required for the ring proof. The result for algebras is

4.4.1   THEOREM   *A nilpotent algebra A contains only one subalgebra S of some given finite dimension, $0 \neq S \neq A$, if and only if one of the following conditions holds:*
(1) dim $S =$ dim $A - 1$ *and A is a power algebra.*
(2) dim $S = 1$, dim $A \geq 3$, $\mathfrak{A}(A) = A^2 = S$, *and $x \in A$, $x^2 = 0$ implies $x \in S$.*

Remark 1   Note that the algebras of case (2) have the property that every subalgebra is an ideal. In §6 we shall find that the structure of an arbitrary algebra in which every subalgebra is an ideal is closely related to the structure of algebras satisfying (2). These algebras seem in one way analogous to the quaternion group of order 8, which plays a key role both in the determination of p-groups with a unique subgroup of order p, and in the determination of groups in which all subgroups are normal.

Remark 2   The classification of the finite-dimensional algebras A over a field F satisfying (2) of **4.4.1** is closely related to the study of quadratic forms over F. Let A have a basis $\{a_1, a_2, \ldots, a_n, b\}$ with $b \in A^2$, and choose $\alpha_{ij} \in F$, $1 \le i, j \le n$, so that $a_i a_j = \alpha_{ij} b$. Then condition (2) requires that the quadratic form

$$\sum_{i=1}^{n} \sum_{j=1}^{n} \alpha_{ij} x_i x_j$$

have no non-trivial zero $(x_1, x_2, \ldots, x_n)$. Let us note that when F is a finite field, then every quadratic form in three variables has a non-trivial zero (see, for example, O'Meara [1], p. 158), so if F is finite then dim $A = 3$. On the other hand, over each finite field there exists a quadratic form in two variables with no non-trivial zero, so algebras satisfying (2) always occur when F is finite.

Proof of **4.4.1**   It is easy to check that nilpotent algebras satisfying (1) or (2) have unique subalgebras of the dimensions indicated. For the converse we shall first establish two lemmas.

4.4.2 Lemma   *If A is a nilpotent algebra of dimension 4, over a field F, then A has more than one subalgebra of dimension 2.*

Proof   If dim $A^2 = 3$, then A is a power algebra. Let $A = \langle a \rangle$. Then all subalgebras $\langle a^2 + \varphi a^3, a^4 \rangle$ for different $\varphi \in F$ are distinct and all have dimension 2. If dim $A^2 \le 1$ then A/A^2 is null, and A contains more than one subalgebra of dimension 2. Thus we can suppose that dim $A^2 = 2$, and A^2 is the only subalgebra of A of dimension 2. It follows, for any $x \in A$, $x \notin A^2$, that $x^3 \ne 0$. Since dim $A^2 = 2$, there are elements $a, b \in A$ which are linearly independent mod A^2, so $\{a, b, a^2, a^3\}$ is a basis for A. Choose $\alpha, \beta \in F$ so that $ab = \alpha a^2 + \beta a^3$, and let $b' = b - \alpha a - \beta a^2$. Then $b' \notin A^2$, and $ab' = 0$. Let $b'^2 = \gamma a^2 + \delta a^3$. Then $0 = (ab')b' = ab'^2 = \gamma a^3$ so $\gamma = 0$. Then $b'^3 = 0$ and $b' \notin A^2$, a contradiction.

4.4.3 Lemma   *Let A be a nilpotent algebra with a unique subalgebra S of dimension 1. If $x \in A$, $x \notin S$, then $0 \ne x^2 \in S$.*

Proof   Let $x \in A$, $x \notin S$. Let $e = \exp x$. If $e \ge 4$ then $(x^{e-2})^2 = 0$, so $\langle x^{e-2} \rangle$ has dimension 1, so $\langle x^{e-2} \rangle = S$. But, by **1.3.12**, S is an ideal in A, whence $S \subseteq \mathfrak{A}(A)$, so $xx^{e-2} = 0$, contrary to $e = \exp x$. If $e = 2$ then dim $\langle x \rangle = 1$, contrary to $x \notin S$. Thus $e = 3$. Then dim $\langle x^2 \rangle = 1$, so $\langle x^2 \rangle = S$, so $0 \ne x^2 \in S$.

Proof of **4.4.1**, *continued*   Let A be a nilpotent algebra with a unique subalgebra S of a given dimension, $0 \neq S \neq A$. If dim $A = $ dim $S + 1$, then A is a power algebra, and condition (1) of the conclusion holds. If dim $A \geq$ dim $S + 2$ and dim $S \geq 2$ then by **1.3.12** A contains a subalgebra B with dim $B = $ dim $S + 2$, and an ideal I with dim $I = $ dim $S - 2$. Then the algebra B/I fails to satisfy **4.4.2**. Hence we can suppose dim $S = 1$, dim $A \geq 3$.

Next we show that $A^2 = S$. Choose $x, y \in A$. By **4.4.3** $x^2 \in S$, $y^2 \in S$, and $(x + y)^2 \in S$. Thus $xy + yx \in S$. Then $0 = x(xy + yx) = xyx$, since $x^2 \in S \subseteq \mathfrak{A}(A)$. Thus $(xy)^2 = 0$, so dim $\langle xy \rangle \leq 1$, so $xy \in S$. Thus $A^2 \subseteq S$. $A^2 \neq 0$ is trivial, so $A^2 = S$. That A satisfies (2) of **4.4.1** now follows directly from **4.4.3**. This completes the proof.

We now turn to rings. We shall establish the following

4.4.4 THEOREM *A nilpotent p-ring R contains only one subring S of a given order, $0 \neq S \neq R$, if and only if R and S satisfy one of the following conditions:*

(1) *R is cyclic or quasi-cyclic.*

(2) *$[R{:}S] = p$. R is a power ring.*

(3) *$|S| = p$. Let $A = \{x \in R \mid px = 0\}$. Then A has rank 2 or 3, $A^2 = S$, and $x \in A$, $x^2 = 0$ implies $x \in S$. There is, moreover, an ideal C of R such that $R = C + A$, $C \cap A = S$, and C is cyclic or quasi-cyclic.*

(4) *$|S| = p^2$. $|R| = p^4$, R has type $_p(2, 2)$ and, if $x \in R$ with $px \neq 0$, then $px^2 = 0$ and x^2 is not a natural multiple of x.*

Remark A description of the nilpotent p-rings R satisfying (3) and (4) may be completed in terms of generators and relations as follows:

(3) Let S be generated by an element s. Then $ps = 0$ and $s \in \mathfrak{A}(R)$. If C is quasi-cyclic, then $C \subseteq \mathfrak{A}(R)$. The subring A satisfies one of the following conditions:

(a) A has a basis $\{s, b\}$, and $b^2 \neq 0$.

(b) A has a basis $\{s, b_1, b_2\}$. Let $b_i b_j = B_{ij} s$, $i, j = 1, 2$, $0 \leq B_{ij} < p$. Then $B_{12} X^2 + (B_{12} + B_{21}) XY + B_{22} Y^2 \equiv 0 \pmod{p}$ for integers X and Y implies $X \equiv Y \equiv 0 \pmod{p}$.

(4) Let R have a basis $\{a_1, a_2\}$. Let $a_i a_j = A_{ij} p a_1 + B_{ij} p a_2$, $i, j = 1, 2$. Then

$$B_{11} X^3 + (A_{11} + B_{12} + B_{21}) X^2 + (B_{22} + A_{12} + A_{21}) X + A_{22} \equiv 0$$

$\pmod{p}$ has no integer solution X.

Since, over the field of p elements, there are both quadratic forms in two variables which do not represent 0 non-trivially, and irreducible cubic polynomials, rings satisfying (3) and (4) occur nontrivially for all primes p.

Proof of **4.4.4**   It is easy to see that nilpotent p-rings satisfying
$(1) - (4)$ have unique subrings of the orders indicated. An infinite
nilpotent p-ring which contains only one subring S of a given (finite)
order clearly satisfies one of $(1) - (4)$ if and only if each of its finite
subrings which properly contains S also does. Thus for the converse
we consider only finite rings. As a notational convenience, let $U(n, s)$
denote the class of nilpotent rings of order p^n which contain only one
subring, generically denoted S, of order p^s. If $R \in U(n, n - 1)$, then
the basis theorem **4.1.4** implies R is a power ring. The rings in $U(n, 1)$
are studied in §5. To characterize the rings in $U(n, s)$, $1 < s < n - 1$,
we first determine those in $U(4, 2)$, $U(5, 2)$, and $U(5, 3)$ and then
proceed by induction. Several steps of the proof are separated as
lemmas.

4.4.5   *Let R be a nilpotent ring of type $_p(n, 1)$. Then, for $1 < i < n$,
R has exactly $p + 1$ ideals of order p^i.*

Proof   For $1 < i \leq n + 1$, $B_i = \{x \in R \mid p^{i-1}x = 0\}$ is an ideal of
R of order p^i. For $1 \leq i < n$, $C_i = \{p^{n-i}x \mid x \in R\}$ is an ideal of R of
order p^i. Hence R has at least two, so that by the Anzahl theorem
4.3.5 at least $p + 1$, ideals of order p^i, $1 < i < n$. But these exhaust
the subgroups of R^+ of order p^i.

4.4.6   *Suppose R is a power ring, $R \in U(n, s)$, $1 \leq s < n - 1$. Then
R is cyclic.*

Proof   First suppose $s = 1$. Let R be generated by an element a,
and let $\Phi = pR + R^2$. Let $M = \{x \in \Phi \mid px = 0, x \notin S\}$. If M is non-
empty, then by **1.2.2** there is some $d \in M$ such that $da \in S$. From
$d \in \Phi$ follows $d = py + az$, some $y, z \in R$. Then, since $pd = 0$ and
$S \subseteq \mathfrak{A}(R)$, $d^2 = d(py + az) = 0$, so d generates a second subring of
order p. Thus M is empty and Φ is cyclic. Now $[R:\Phi] = p$, since R is
a power ring. If R had type $(n - 1, 1)$, then $\Phi = \{x \in R \mid p^{n-2}x = 0\}$,
so Φ would not be cyclic. Hence R is cyclic, as desired.

We now proceed by induction on s. Suppose $s > 1$. Let I be an ideal
of order p of R. Applying the induction hypothesis to the power ring
R/I we find that R/I is cyclic. Hence either R is cyclic or has type
$(n - 1, 1)$. But type $(n - 1, 1)$ is excluded by **4.4.5**.

Remark   It may be amusing to note the following corollary to **4.4.6**:

*Let R be a finite nilpotent p-ring, $\mid R \mid \neq 4$. If the circle group of R is
cyclic then R is cyclic.*

The proof is direct if $\mid R \mid \leq p^2$, $\mid R \mid \neq 4$. If $\mid R \mid > p^2$ observe that every subring of R is a subgroup of (R, o), so R has only one subring of each order, and hence satisfies **4.4.6**. This corollary is proved independently by Eldridge and Fischer [2].

4.4.7   *If $R \in U(4, 2)$, then the rank of R is at most* 2.

Proof   Suppose $R \in U(4, 2)$. Lemma **4.4.2** for $F = GF(p)$ excludes rank $R = 4$. Suppose rank $R = 3$, so R has type $_p(2, 1, 1)$. Let $T = \{x \in R \mid px = 0\}$. Since $\mid T \mid = p^3$, T contains S, the unique subring of order p^2. Thus T is a power p-algebra. It follows that R has a basis of the form $\{x, y, y^2\}$ where char $x = p^2$, char $y = $ char $y^2 = p$, and $y^3 = px$. Since $\mid \langle y^2, y^3\rangle \mid = p^2$, $\mid R^2 \mid \geq p^2$. By **4.4.6** R is not a power ring, so $\mid R^2 \mid \leq p^2$. Thus $R^2 = S = \langle y^2, y^3\rangle$, so there are integers A, B such that $xy = Ay^2 + By^3$. Let $x' = x - Ay - By^2$. Then $x'y = 0$. Let $x'^2 = Cy^2 + Dy^3$ for integers C, D. Then
$$0 = x'(x'y) = x'^2y = Cy^3$$
so $C \equiv 0 \pmod{p}$. Thus $x'^2 = Dy^3 = Dpx'$, so $\langle x'\rangle \neq S$ is a second subring of order p^2. Thus rank $R \leq 2$.

4.4.8   *If $R \in U(5, s)$, $s = 2, 3$, then R is cyclic.*

Proof   Suppose rank $R \geq 3$. Then, for $s = 2$ (resp. for $s = 3$), we can find a subring A of order p^4 (resp. an ideal I of index p^4) with rank $A \geq 3$ (resp. rank $(R/I) \geq 3$). This is impossible by **4.4.7**, so rank $R \leq 2$.

Suppose rank $R = 2$. By **4.4.5** R has type $(3, 2)$. Let $\{a, b\}$ be a basis for R, char $a = p^3$, char $b = p^2$. Since pa and $\{p^2a, pb\}$ generate distinct subrings of order p^2, we have $s = 3$. Then $pR = S$. Moreover, $R^2 = S$, since $R^2 \subset S$ or $R^2 \supset S$ implies R is a power ring, which is excluded by **4.4.6**. Let $a^2 = Apa + Bpb$, $b^2 = Cpa + Dpb$, $C \not\equiv 0$ $\pmod{p}$, $ab = Epa + Fpb$. By replacing a by $a' = a - EC^{-1}b$, where $C^{-1}C \equiv 1 \pmod{p^2}$, we may assume that $E = 0$. Then $(a^2)b = BCp^2a$, while $a(ab) = 0$. Thus $B \equiv 0 \pmod{p}$, and a generates a second subring of order p^3.

Proof of **4.4.4**, *continued*   If $R \in U(4, 2)$, then by **4.4.5** and **4.4.7** either R is cyclic or has type $(2, 2)$. If R has type $(2, 2)$ and, for some $x \in R$, $px \neq 0$ and x^2 is a multiple of x, then both pR and $\langle x\rangle$ have order p^2. Thus (4) of **4.4.4** holds.

Suppose $R \in U(n, s)$ with $n > 5$, $1 < s < n - 1$. If $s = 2$ and rank $R \geq 2$, we can find a subring of order p^5 and rank ≥ 2, which contradicts **4.4.8**. For $s > 2$ we proceed by induction on n. Let I be an ideal of R of order p. By induction hypothesis R/I is cyclic. R cannot have type $(n-1, 1)$ by **4.4.5**, hence R is cyclic.

5 Nilpotent p-rings with one subring of order p

In this section we shall show that a finite nilpotent p-ring R which contains a unique subring S of order p satisfies condition (3) of Theorem **4.4.4**. Let S be generated by an element s. Then $ps = 0$ and $sR = Rs = 0$. The letters x and y will denote elements of R. For ease of reference we restate the hypothesis that S is the only subring of order p.

4.5.1 *If $px = x^2 = 0$, then $x \in S$.*

4.5.2 *Suppose char $x = p^a$ and $p^{a-1}x \notin S$, $a \geq 1$. Then $a = 1$ and $x^2 \in S$, $x^2 \neq 0$.*

Proof By **4.5.1**, $(p^{a-1}x)^2 \neq 0$. This, with $p^a x^2 = 0$, gives $2a - 2 < a$, so $a = 1$. Then $\langle x, S \rangle$ is a p-algebra, so **4.4.3** implies $x^2 \in S$, $x^2 \neq 0$.

4.5.3 LEMMA *Let a_1, a_2, a_3, and b be elements of a ring such that $pb = 0$ and there are integers A_{ij}, $0 \leq A_{ij} < p$, $i,j = 1, 2, 3$, such that $a_i a_j = A_{ij}b$. Then there exist integers $0 \leq X_i < p$, $i = 1, 2, 3$, not all 0, such that*

$$(X_1 a_1 + X_2 a_2 + X_3 a_3)^2 = 0.$$

Proof This is equivalent to the well-known fact that a quadratic form in three variables over the field of p elements represents 0 non-trivially. (See, for example, O. T. O'Meara [1], p. 158.)

4.5.4 LEMMA *Let $A = \{x \in R \mid px = 0\}$. Then $A^2 \subseteq S$ and one of the following conditions holds, according to the rank of A^+:*
(1) *$A = S$*
(2) *A has a basis $\{s, b\}$, and $b^2 \neq 0$.*
(3) *A has a basis $\{s, b, c\}$, and $(Xb + Yc)^2 = 0$ for integers X and Y implies $X \equiv Y \equiv 0 \pmod{p}$.*

Proof Surely $S \subseteq A$, so A is a p-algebra with a unique subalgebra of dimension 1. The result follows directly from **4.4.1** and **4.5.3**.

From now on the letters b, and, when appropriate, c, denote members of a fixed basis for A.

In case $pR = 0$ we have $R = A$, and thus R satisfies (3) of **4.4.4**.

If $pR \neq 0$, then, by **4.5.2**, $R^+ = C^+ + A^+$ where C^+ is a cyclic p-group with $|C^+| > p$, and $C^+ \cap A^+ = S^+$. The rest of the proof is devoted to showing that $(R^2)^+ \subseteq p\,C^+$. This implies that the set of elements of C^+ forms a subring C, and thus (3) of **4.4.4** holds. Let a be a generator of C^+. The proof that $(R^2)^+ \subseteq pC^+$ is divided into cases depending on the location of a^2 and on the rank of A, which of course equals the rank of R.

If A has rank 1, then $C = R$ so (1) of **4.4.4** holds. Suppose A has rank 2, with basis s, b. If $a^2 \notin pC^+$, then R^2 has rank 2 so $[R:\Phi_R] = p$, which contradicts **4.4.6**. Thus $a^2 \in pC^+$. By **4.5.4** $A^2 \subseteq S$, and $S^+ \subseteq pC^+$. Finally, $ab \in S$ and $ba \in S$ by the nilpotence of b.

Thus we may assume that A has rank 3, with basis s, b, c. If R^2 has rank 1, we are done. If R^2 has rank 3, then R is a power ring, contrary to **4.4.6**. Thus assume R^2 has rank 2. Without loss of generality we may assume $b \in R^2$. To complete the proof we make use of the following remark:

4.5.5   *Under the above assumptions, if $x \in R^2$, $px = 0$, and $xb = 0$, then $x \in S$.*

Proof   Since R^2 has rank 2 and s, $b \in R^2$, it follows that $x = Xs + Yb$, some integers X and Y. Since $s \in \mathfrak{A}(R)$, $xb = Yb^2$. Since $b^2 \neq 0$ but $xb = 0$, $Y \equiv 0 \pmod{p}$. Thus $x \in S$.

We now continue the proof. Since $b^2 \in S$, $0 = ab^2 = (ab)b$ so by **4.5.5** $ab \in S$. Dually $ba \in S$. By **4.5.4** $A^2 \subseteq S$. Since $cb \in S$,
$$0 = a(cb) = (ac)b$$
so, by **4.5.5**, $ac \in S$. Dually, $ca \in S$. Since $ab \in S$, $0 = a(ab) = a^2b$. Thus $R^2b = 0$. Choose any $x \in R^2$. Since R^2 has rank 2, and $b \in R^2$, $S \subseteq R^2$, we can write $x = pX_1a + X_2b$ for some integers X_1 and X_2. Then $0 = xb = (pX_1a + X_2b)b = X_2b^2$. Since $b^2 \neq 0$, $X_2 \equiv 0 \pmod{p}$. Thus $x \in pC^+$, so that R^2 is cyclic.

6 Rings in which all subrings are ideals

This section studies several properties of rings in which all subrings are two-sided ideals, which in analogy with Hamiltonian groups are called *Hamiltonian rings*. For convenience we abbreviate the term Hamiltonian ring [p-ring] to H-*ring* [H-p-*ring*].

It is easy to see that the direct sum of H-p-rings for different primes p is an H-ring. Hence to find the structure of H-rings with no elements of characteristic 0 it will be sufficient for us to study H-p-rings. The

structure of H-rings which contain elements of characteristic 0 is
slightly more complicated and will not be discussed here. A description
of this structure appears in Kruse [2]. In this section we first obtain
a result by which the study of H-p-rings is reduced to that of nil
H-p-rings. Second, we introduce a class of p-rings called "almost-null"
and obtain several of their properties, showing that a nil p-ring of
characteristic 0 is an H-ring if and only if it is almost-null. A corollary
is that every nil H-p-ring is nilpotent. The structure of almost-null
p-rings is closely related to that of nilpotent p-rings with a unique
subring of order p. Third, we prove one of the main results of this
section. if R is a nil H p ring, then R^2 is finite, and has rank at most 2.
We conclude the section by determining all nilpotent H-p-rings R for
which R^2 is cyclic. Those nilpotent H-p-rings R for which R^2 has
rank 2 are more complicated, but must decompose as a direct sum
$R = S \oplus N$ where N is a null p-ring and S has rank at most 4. The
enumeration of possible defining relations for the subring S are rather
lengthy, although relatively straightforward. For the details see Kruse
[1, 3].

In the sequel we shall make frequent use of the simple observations
that (1) all cyclic rings and all null rings are H-rings; (2) subrings and
homomorphic images of H-rings are again H-rings; and (3) a ring is
an H-ring if and only if every subring generated by a single element is
an ideal. In this section only we shall use the symbols $\{S\}$ to denote
the additive subgroup generated by a subset S of a ring, and $S \dotplus T$
to denote an additive group direct sum of subgroups S and T of a ring.
The meanings of the symbols $\langle S \rangle$ and $S \oplus T$ remain unchanged.

4.6.1 THEOREM *A ring R is an H-p-ring if and only if R satisfies
one of the following conditions*:
(1) *R is a nil H-p-ring.*
(2) *$R \cong GF(p) \oplus N$ where N is a nil H-p-ring.*
(3) *$R \cong Z/\langle p^n \rangle$ where Z is the ring of rational integers and n is a
natural number.*

The proof requires two lemmas. The first is a classical result of
Dickson [4, p. 80]. The proof is from Köthe [1].

4.6.2 LEMMA *Let N be a nil ideal of a ring R. If R/N contains an
idempotent, then R contains an idempotent.*

Proof If x is any element of an idempotent coset of R/N, define
$x' = x^2 - x$. Thus x' is always nilpotent. Choose e in an idempotent
coset so that exp e' is minimal. We shall show that $e' = 0$, so that e

is an idempotent. To do this define $y = e + e' - 2ee'$. Then y is in the same coset as e, and $y' = 4e'^3 - 3e'^2$, so $\exp y' < \exp e'$ unless $e' = 0$. This completes the proof.

4.6.3 LEMMA (P. A. Freĭdman [3]) *The radical of an H-p-ring is the set of all nilpotent elements in the ring. A semi-simple H-p-ring is isomorphic to $GF(p)$.*

Proof Let R be an H-p-ring, and let N be the set of all nilpotent elements of R. Since R is an H-ring $xy \in \langle x \rangle$ and $yx \in \langle x \rangle$ for all $x \in N$, $y \in R$. Further, $\langle x \rangle \subseteq N$, so N is an ideal of R. Since N is surely contained in the radical of R, to prove the lemma it is sufficient for us to show that $T = R/N$ is isomorphic to $GF(p)$. Certainly $pR \subseteq N$, so T is a p-algebra. Choose $x \in T$, $x \neq 0$. Since T is an H-ring, $\langle x^2 \rangle$ is an ideal in T, so $xx^2 = x^3 = F(x^2)$ for some polynomial F over $GF(p)$. Solving $x^3 - F(x^2) = 0$ for the highest power of x, we obtain that $\langle x \rangle$ is a finite-dimensional, non-nilpotent algebra. By a classical result (Peirce [1], p. 113), $\langle x \rangle$ contains an idempotent e. The Peirce decomposition gives $T = \langle e \rangle \oplus S$, where $S = \langle x \in T \mid ex = 0 \rangle$. As a direct summand of a semi-simple ring, S is semi-simple. If $S \neq 0$ then, as above, S contains an idempotent f. Then $e(e + f) \notin \langle e + f \rangle$, which contradicts the assumption that T is an H-ring. Thus $S = 0$, so $T = \langle e \rangle \cong GF(p)$. This completes the proof.

Proof of **4.6.1** Rings satisfying (1) and (3) are evidently H-p-rings. To show that $GF(p) \oplus N$ is an H-ring, where N is a nil H-p-ring, it is sufficient to show that each subring $\langle f + x \rangle$ is an ideal, $f \in GF(p)$, $x \in N$. Let $r = \exp x$. Then $(f + x)^r = f^r$ and $\langle f \rangle = \langle f^r \rangle$, so
$$\langle f + x \rangle = \langle f \rangle + \langle x \rangle$$
is the sum of ideals of direct summands, and so is an ideal in the direct sum.

Now suppose R is an H-p-ring which is not nil, and let N be the radical of R. By **4.6.3** R/N contains an idempotent, and N is nil. By **4.6.2**, then, R contains an idempotent e. The Peirce decomposition gives $R = \langle e \rangle \oplus M$ where $M = \langle x \in R \mid ex = 0 \rangle$. If M were not nil it would, as above, contain an idempotent, and R/N would then fail to satisfy **4.6.3**. Thus $M = N$ is nil.

Let $p^n = \operatorname{char} e$. If $n = 1$ then R satisfies (2) of the conclusion. For $n > 1$ the inclusion
$$e(pe + x) = pe \in \langle pe + x \rangle, x \in N,$$
implies $x = 0$. Thus $N = 0$, so R satisfies (3) of the conclusion.

We now turn to the structure of nil rings in which all subrings are ideals, and begin by defining a property of fundamental importance.

4.6.4 DEFINITION An additive subgroup S of a p-ring R *almost annihilates* R provided, for all $x \in S$, $x^3 = px^2 = 0$, and $Rx + xR \subseteq \{x^2\}$. In the case when $R = S$, the ring R is called *almost-null*.

Observe that when R is a nil H-p-ring, then the condition $x^3 = 0$ follows from $px^2 = 0$ and $Rx + xR \subseteq \{x^2\}$. Further, if S almost annihilates R and $x \in S$, then x annihilates R if and only if $x^2 = 0$. To show the importance of almost-null rings we shall obtain the following

4.6.5 THEOREM *A nil p-ring of characteristic 0 is an H-ring if and only if it is almost-null.*

The proof requires several lemmas.

4.6.6 LEMMA *If x is an element of a nil H-p-ring, then $x^3 \in \{x^2\}$.*

Proof Let k be the minimal integer such that $x^k = mx^{k-1}$ for some integer m. Suppose $k \geq 4$, and let $e = \exp x$. Then $0 = x^{e-k}x^k = mx^{e-1}$ implies p divides m. From $xx^2 \in \langle x^2 \rangle$ it follows that $x^3 = F(x^2)$ for some polynomial with integral coefficients $\alpha_1, \ldots, \alpha_r$, so

$$x^3 = F(x^2) = \alpha_1 x^2 + \alpha_2 x^4 + \ldots + \alpha_r x^{2r},$$

so $x^{k-1} = x^{k-4}x^3 = x^{k-4}F(x^2) = \alpha_1 x^{k-2} + (\alpha_2 m + \ldots + \alpha_r m^{2r-3})x^{k-1}$. Since p divides m and char x is a power of p, this equation may be solved for x^{k-1} as an integral multiple of x^{k-2}. This contradicts the minimality of k. Thus $k \leq 3$, so the conclusion holds.

4.6.7 COROLLARY *Let R be a nil H-p-ring and let $x \in R$, $y \in R$. Then there are integers α, β, γ, δ such that*

$$xy = \alpha x + \beta x^2 = \gamma y + \delta y^2.$$

4.6.8 LEMMA *Let x and y be elements of a nil H-p-ring such that char $y \geq p^{2m}$, where $p^m = $ char x. Then $px^2 = 0$.*

Proof By **4.6.7**, $x(px) = \alpha px + \beta p^2 x^2$ for suitable integers α, β. It follows that $px^2 \in \{x\}$. First suppose $\{x\} \cap \{y\} = 0$. Then

$$x(px + p^m y) \in \langle px + p^m y \rangle$$

implies $px^2 = 0$, q.e.d. Second suppose $\{x\} \cap \{y\} \neq 0$. Let $p^r = $ char y. Then $r \geq 2m$, and there are integers $\alpha \not\equiv 0 \pmod{p}$ and s, $0 \leq s \leq m$, such that $p^{r-m+s}y = \alpha\, p^s x$. and $\{x\} \cap \{y\} = \{p^s x\}$. Then

$$x(\alpha px - p^{r-m+1}y) \in \langle \alpha px - p^{r-m+1}y \rangle$$

implies $px^2 = 0$. This completes the proof.

Proof of **4.6.5** An almost-null p-ring is clearly a nil H-p-ring. For the converse suppose that R is a nil H-p-ring of characteristic 0. By **4.6.8**, $px^2 = 0$ for all $x \in R$. Thus to show that R is almost-null it is sufficient to show that $xy \in \{x^2\} \cap \{y^2\}$ for all x, $y \in R$. We shall show that $xy \in \{x^2\}$. $xy \in \{y^2\}$ follows dually.

Let char $x = p^r$, char $y = p^s$, and define $t = \max(r, s) + 1$. Choose $z \in R$ so that char $z = p^{2t}$. Then $xy \in \{x^2\}$ follows from $wy \in \langle w \rangle$, where $w \in R$ is chosen as follows:

(1) If $\{z\} \cap \langle x \rangle = 0$ then choose $w = x + p^t z$.

(2) If $\{z\} \cap \{x\} = \{p^k x\} \neq 0$ then $p^{2t+k-r}z = \alpha p^k x$ for some integer $\alpha \not\equiv 0 \pmod p$. Choose $w = \alpha x - p^{2t-r}z$.

(3) If $\{z\} \cap \langle x \rangle \neq 0$ but $\{z\} \cap \{x\} = 0$, then there exist integers α and β such that $p^{2t-1}z = \alpha p^{r-1}x + \beta x^2$. If $\alpha \equiv 0 \pmod p$ then choose $w = x + p^t z$. If $\alpha \not\equiv 0 \pmod p$ then choose $w = \alpha x - p^{2t-r}z$.

This completes the proof.

4.6.9 COROLLARY *Every nil H-p-ring is nilpotent.*

Proof Let R be a nil H-p-ring. If char $R = 0$ then **4.6.5** and **4.6.4** imply that $R^3 = 0$. Suppose char $R \neq 0$. Then the ideal pR is nilpotent, so it is sufficient to show that the nil p-algebra $T = R/pR$ is nilpotent. By **4.6.6** $x^3 = 0$ for all $x \in T$. Choose x, y, $z \in T$. By the nilpotence of x, $xy \in \langle x^2 \rangle$ and $xz \in \langle x^2 \rangle$. Thus $xyz = \alpha x^2 z = \alpha\beta x^3 = 0$ for suitable integers α and β. Thus $T^3 = 0$.

Next we determine the structure of almost-null p-rings.

4.6.10 THEOREM *A p-ring R is almost-null if and only if R satisfies one of the following conditions:*

(1) *R is null.*

(2) *$R = \{x\} + \mathfrak{A}(R)$ where $x^2 \in \mathfrak{A}(R)$, $px \in \mathfrak{A}(R)$, and char $x^2 = p$.*

(3) *$R = \{x_1, x_2\} + \mathfrak{A}(R)$, $px_1 \in \mathfrak{A}(R)$, $px_2 \in \mathfrak{A}(R)$, and there exist $z \in \mathfrak{A}(R)$, char $z = p$, and integers α_{ij} such that $x_i x_j = \alpha_{ij} z$, i, $j = 1, 2$, and such that the congruence*

$$\alpha_{11}\varphi^2 + (\alpha_{12} + \alpha_{21})\,\varphi\psi + \alpha_{22}\psi^2 \equiv 0 \pmod p \qquad (*)$$

for integers φ and ψ implies $\varphi \equiv \psi \equiv 0 \pmod p$.

Proof A straightforward verification establishes that the p-rings R described in (1), (2), and (3) are almost-null. For the converse suppose R is an almost-null p-ring. Since $pR^2 = 0$, **1.2.5** implies $pR \subseteq \mathfrak{A}(R)$, so that $R/\mathfrak{A}(R)$ is a p-algebra. Let $n = \dim (R/\mathfrak{A}(R))$. An element $x \in R$, moreover, satisfies $x^2 = 0$ if and only if $x \in \mathfrak{A}(R)$. The cases $n = 0$ and $n = 1$ thus lead directly to cases (1) and (2) of the conclusion. Suppose $n \geq 2$. We shall show that $\mid R^2 \mid = p$. For choose $x, y \in R$, $x, y \notin \mathfrak{A}(R)$. If $xy \neq 0$ or $yx \neq 0$ then $\{x^2\} = \{y^2\}$ follows from the hypothesis that $xy, yx \in \{x^2\} \cap \{y^2\}$. If $xy = yx = 0$ then

$$x^2 = x(x + y) \in \{(x + y)^2\}$$

implies $\{x^2\} = \{y^2\}$. Thus all squares in R, and so by the definition of almost-null all products, are multiples a fixed element $z \in \mathfrak{A}(R)$, char $z = p$. It now follows from **4.5.3** that $n = \dim (R/\mathfrak{A}(R)) \leq 2$. Thus $n = 2$. Choose $x_1, x_2 \in R$, linearly independent mod $\mathfrak{A}(R)$, and let $x_i x_j = \alpha_{ij} z$, $i, j = 1, 2$. The condition that $(\varphi x_1 + \psi x_2)^2 \neq 0$ unless $\varphi \equiv \psi \equiv 0 \pmod{p}$ is equivalent to the condition (*) of the conclusion. This completes the proof.

Remarks on algebras An algebra in which all subalgebras are ideals is called an H-*algebra*. Forming the obvious analogies with the preceding results in this section one obtains:

Let A be an algebra over a field F. Then A is an H-algebra if and only if (1) A is a nil H-algebra, (2) $A \cong F$, or (3) $A \cong F \oplus N$ where N is a nil H-algebra over F.
A nil algebra A is an H-algebra if and only if A is almost-null, where an almost-null algebra N is defined by the properties that $x^3 = 0$ for all $x \in N$, and $xy, yx \in \langle x^2 \rangle \cap \langle y^2 \rangle$ for all $x, y \in N$.
An algebra A is almost-null if and only if A has a vector space direct sum decomposition $A = B \dotplus \mathfrak{A}(A)$ such that $x \in B$, $x^2 = 0$ implies $x = 0$, and there exists $z \in \mathfrak{A}(A)$ such that $A^2 = B^2 = \langle z \rangle$.

Note that the algebras $\langle B, z \rangle$ have unique subalgebras of dimension 1. As was mentioned in §4, these algebras seem in one way analogous to the quaternion group of order 8, which plays a key role both in the classification of Hamiltonian groups and of p-groups with a unique subgroup of order p. The classification of the algebras $\langle B, z \rangle$ is closely related to the study of quadratic forms over the field. See Remark 2 following **4.4.1**.

The classification of H-algebras was done by Liu [1]. These results have been extended to certain classes of non-associative algebras by Outcalt [1].

F

We now return to the study of nil H-p-rings. Those which contain elements of arbitrarily large characteristic are completely characterized by Theorems **4.6.5** and **4.6.10**. Every abelian p-group, however, which does not possess elements of arbitrarily large order may be decomposed as a direct sum of cyclic subgroups (see, e.g., Fuchs [1], p. 43). Hence, whenever convenient in the sequel, we shall, without loss of generality, assume that all p-rings have (additive group) bases. We now state one of the main results of this section.

4.6.11 THEOREM   *If R is a nil H-p-ring, then R^2 has rank at most 2.*

The proof requires several lemmas.

4.6.12 LEMMA   *A nil p-ring generated by a single element x is an H-ring if and only if one of the following conditions holds:*
I. $x^2 \in \{px\}$ *(so $\langle x \rangle$ is cyclic).*
II. $x^2 \notin \{px\}$ *but $px^2 \in \{px\}$, $x^3 \in \{px\}$.*

An element x of a nil H-p-ring is called *type I* or *type II* according as it satisfies I or II of **4.6.12**.

Proof   Consider the inclusion $x(px) \in \langle px \rangle$. By **4.6.7** this gives $px^2 = \alpha px + \beta p^2 x^2$ for suitable integers α, β. Thus $p(1 - \beta p)x^2 \in \{px\}$ Since $1 - \beta p \not\equiv 0 \pmod p$, either $x^2 \in \{px\}$, so that case I of the conclusion holds, or else $x^2 \notin \{px\}$ but $px^2 \in \{px\}$. By **4.6.6**, $x^3 \in \{x^2\}$, and then, by the nilpotence of x, $x^3 \in \{px^2\} \subseteq \{px\}$. Thus case II holds. The proof of the converse is straightforward.

4.6.13 COROLLARY   *If x and y are elements of a nil H-p-ring, then $p(xy) \in \{x\} \cap \{y\}$. If $\{x\} \cap \{y\} = 0$ then $p(xy) = p(yx) = 0$.*

A major method used in the proofs below is to consider the implications of inclusions of the form $xw \in \langle w \rangle$, where x is a member of a basis for a nil H-p-ring, and w is expressed as an integer linear combination of basis elements. Since products of distinct basis elements have characteristic at most p, the implications of $xw \in \langle w \rangle$ are generally obvious after one has found a basis for the p-algebra of elements of characteristic p in $\langle w \rangle$. If w is type I then the algebra is one dimensional. If w is type II let $pw^2 = mp^r w$ and char $w = p^s$, where m is an integer, $m \not\equiv 0 \pmod p$. The algebra is two dimensional, with a basis $w^2 - mp^{r-1}w$ and $p^{s-1}w$. Since the details of finding the basis for the algebra in $\langle w \rangle$ are straightforward, they will be omitted.

4.6.14 LEMMA   *Let x and y be elements of a nil H-p-ring such that $\{x\} \cap \{y\} = 0$ and char $x \geq$ char y. Then $py^2 = 0$.*

Proof By **4.6.12** $px^2 \in \{x\}$ and $py^2 \in \{y\}$. Let r and s be the maximal integers, $p^r \leq$ char x and $p^s \leq$ char y, such that $px^2 \in \{p^r x\}$ and $py^2 \in \{p^s y\}$. Suppose $py^2 \neq 0$. Then $y(px + py) \in \langle px + py \rangle$ implies $r < s$. Then $x(px + py) \notin \langle px + py \rangle$. Thus $py^2 = 0$.

4.6.15 LEMMA *Let x and y be elements of a nil H-p-ring, $p \neq 2$, such that $\{x\} \cap \{y\} = 0$, x is type II, and char $x \geq$ char y. Then $y^2 \in \langle x \rangle$.*

Remark **4.6.15** fails to hold for $p = 2$. If the remaining hypotheses of **4.6.15** hold, but $p = 2$ and $y^2 \notin \langle x \rangle$, then one can show that char $x =$ char y and $x^2 \in \langle y \rangle$.

Proof Suppose the hypotheses of the Lemma hold, but $y^2 \notin \langle x \rangle$. Define $p^t =$ char y and let $r > 0$, $s \geq 0$, and $m \not\equiv 0 \pmod p$ be those integers such that $px^2 = mp^r x$ and char $x = p^{r+s}$. Note that by **4.6.14** $py^2 = 0$.

First suppose that y is type I, so that $y^2 \in \{y\}$. Then $y^2 \notin \langle x \rangle$ implies $\langle x \rangle \cap \langle y \rangle = 0$. If $r \geq t$ then $x(x + y) \notin \langle x + y \rangle$. If $r < t$ then char $x \geq$ char y implies $s \geq 1$, and then $x(p^s x + y) \notin \langle p^s x + y \rangle$.

Now suppose that y is type II. If $\langle x \rangle \cap \langle y \rangle = 0$ then
$$x(x + y) \notin \langle x + y \rangle.$$
Thus $\langle x \rangle \cap \langle y \rangle \neq 0$, which, with $y^2 \notin \langle x \rangle$, gives
$$p^{t-1} y + cy^2 = a(x^2 - mp^{r-1}x) + bp^{r+s-1}x$$
for suitable integers a, b, c. Let $z = p^{t-1}y + cy^2$. Then for some integers d and e we have $xy = dz$ and $yx = ez$. Suppose $r > t$. Then $x(x + y) \in \langle x + y \rangle \cap \langle x \rangle$ implies $x^2 + dz \in \{p^{r-1}x\}$. Similarly
$$x(x - y) \in \langle x - y \rangle \cap \langle x \rangle$$
implies $x^2 - dz \in \{p^{r-1}x\}$.

Since $p \neq 2$ these conditions together imply $x^2 \in \{p^{r-1}x\}$, contradicting the assumption that x is type II. Thus $r \leq t$. char $x \geq$ char y implies $s \geq t - r$. If $r < t$ then $s > 0$, and $x(px + py) \notin \langle px + py \rangle$. Thus $r = t$. Choose an integer $\varphi \not\equiv 0 \pmod p$ and consider the inclusion $(\varphi x + y)y \in \langle y \rangle \cap \langle \varphi x + y \rangle$. This implies
$$\varphi dz + y^2 \in \{[\varphi^2 + f\varphi + a\varphi(d + e)]z + [a - cf\varphi]y^2\} \qquad (1)$$
where $f = bp^s - am$. By hypothesis z and y^2 are linearly independent, so that the determinant
$$\begin{vmatrix} 1 & a - cf\varphi \\ \varphi d & \varphi^2 + f\varphi + a\varphi(d + e) \end{vmatrix} = [1 + cdf]\varphi^2 + [f + ae]\varphi \equiv 0 \pmod p. \qquad (2)$$

Since $p \neq 2$ and (2) holds for all $\varphi \not\equiv 0 \pmod{p}$, it follows that $1 + cdf \equiv f + ae \equiv 0 \pmod{p}$. By the dual relation

$$y(\varphi x + y) \in \langle y \rangle \cap \langle \varphi x + y \rangle$$

we obtain $1 + cef \equiv f + ad \equiv 0 \pmod{p}$. Thus $d \equiv e \not\equiv 0 \pmod{p}$ and $a \not\equiv 0 \pmod{p}$. Then (1) fails for $\varphi = -ad$. Hence in every case $y^2 \notin \langle x \rangle$ leads to a contradiction, so the proof is complete.

Proof of **4.6.11**   The proof will be given only for $p \neq 2$. The proof for $p = 2$ requires lengthy but straightforward computation, the details of which may be found in Kruse [1]. Let R be a nil H-p-ring, $p \neq 2$. If char $R = 0$ then **4.6.5** and **4.6.10** imply $|R^2| \leq p$, so suppose that char $R \neq 0$. Then R has an additive group basis. To show that rank $R^2 \leq 2$ it is sufficient, for $T = R/pR^2$, to show that rank $T^2 \leq 2$. Let I be an index set, and x_i elements of T, $i \in I$, which form a basis for T. Define $S = \{x_i^2 \mid i \in I\}$.

First suppose that S has rank 2 or less. Then the basis may be chosen so that there are indices k, $l \in I$ such that $S \subseteq \{x_k, x_l\}$. Observe, for $i \in I$, that $\langle x_i \rangle \subseteq \{x_i, x_k, x_l\}$, so that, for $i, j \in I$, $i \neq j$,

$$x_i x_j \in \langle x_i \rangle \cap \langle x_j \rangle \subseteq \{x_i, x_k, x_l\} \cap \{x_j, x_k, x_l\} = \{x_k, x_l\}.$$

Hence $T^2 \subseteq \{x_k, x_l\}$, so T^2 has rank at most 2.

Now suppose that S has rank 3 or more. Then T contains basis elements, x, y, z such that x^2, y^2, and z^2 are linearly independent. Rename x, y, z so that char $x \geq$ char $y \geq$ char z. If x is type II then **4.6.15** implies $y^2 \in \langle x \rangle$ and $z^2 \in \langle x \rangle$. Hence $\{x^2, y^2, z^2\} \subseteq \langle x \rangle$ which, by **4.6.12** has rank at most 2. Thus x is type I.

Suppose $\langle x \rangle \cap \langle y \rangle = 0$. Then at least one of the elements $x + y$ and $x - y$ is type II. Let w be one of these of type II. Then char $w =$ char x, so by **4.6.15** $x^2 \in \langle w \rangle$, $y^2 \in \langle w \rangle$, and $z^2 \in \langle w \rangle$. Hence as above T^2 has rank at most 2.

Finally suppose $\langle x \rangle \cap \langle y \rangle \neq 0$. This, with x of type I and $\{x\} \cap \{y\} = 0$, implies that y is type II and $\langle y \rangle \subseteq \{x, y\}$. Certainly $x^2 \in \{x\} \subseteq \{x, y\}$. By **4.6.15** $z^2 \in \langle y \rangle \subseteq \{x, y\}$. Hence T^2 has rank at most 2. This completes the proof.

Next we turn to the structure of nil H-p-rings R such that R^2 is cyclic. If S is a subgroup of R which almost annihilates R, and $SR + RS = \{x\}$ for some $x \in R$, then we shall say that S almost annihilates R *with respect to* x.

4.6.16   THEOREM   *Let R be a nilpotent p-ring with $|R^2| \leq p$. Then R is an H-ring if and only if*
(1) *R is almost-null; or*

(2) R *contains an element x and a subgroup N such that $R = \{x\} \dotplus N$,
char $N \leq p^k =$ char x, $x^2 = 0$, N almost annihilates R with respect to
$p^{k-1}x$, and char $\mathfrak{A}(R) < p^k$. If, moreover, char $N = p^k$, then $xy = -yx$
for all $y \in R$.*

Proof   It is easy to check that nilpotent p-rings satisfying (1) and
(2) are H-rings. For the converse suppose R is a nilpotent H-p-ring
with $|R^2| = p$, which is not almost-null. By **4.6.4** there exists $x \in R$,
$x \notin \mathfrak{A}(R)$, with $x^2 = 0$. Since $pR \subseteq \mathfrak{A}(R)$, it follows that x has height 0,
and so, when we replace x by $x + y$ for a suitable $y \in pR$ we maintain
$x^2 = 0$, $x \notin \mathfrak{A}(R)$, and obtain a direct sum decomposition $R = \{x\} \dotplus N$
for some subgroup N of R. Since $|R^2| = p$ and $x \notin \mathfrak{A}(R)$, $R^2 \subseteq \langle x \rangle$.
Since $x^2 = 0$, $\langle x \rangle = \{x\}$. Thus $R^2 = \{p^{k-1}x\}$, where $p^k =$ char x.

Choose $y \in N$. If $y^2 \neq 0$ then $\{y^2\} = R^2 = \{p^{k-1}x\}$, so $yR + Ry \subseteq \{y^2\}$.
If $y^2 = 0$ then $yR + Ry \subseteq \langle y \rangle \cap R^2 = \{y\} \cap R^2 = 0$, so that $y \in \mathfrak{A}(R)$.
Thus, by the definition **4.6.4**, N almost annihilates R with respect to
$p^{k-1}x$.

Suppose char $N > p^k$. Then there exists $y \in N$ with char $y > p^k$.
For any such y, $(x + py)y \in \langle x + py \rangle$ implies $xy = 0$. Dually $yx = 0$.
Since $x^2 = 0$ but $x \notin \mathfrak{A}(R)$, there exists $z \in N$ with $xz \neq 0$ or $zx \neq 0$.
It follows that char $z \leq p^k$. Then char $(y + z) > p^k$ and $x(y + z) \neq 0$
or $(y + z)x \neq 0$, a contradiction. Thus char $N \leq p^k$.

Suppose $y \in \mathfrak{A}(R)$ and char $y = p^k$. Choose $z \in N$ so that $xz \neq 0$ or
$zx \neq 0$. Then $(x + y)z \notin \langle x + y \rangle$ or $z(x + y) \notin \langle x + y \rangle$. Thus char $\mathfrak{A}(R)$
$< p^k$.

To complete the proof that R satisfies (2) of the conclusion we assume
that char $N = p^k$ and prove that $xy = -yx$ for all $y \in N$. First suppose
char $y = p^k$. Let $y^2 = ap^{k-1}x$, $xy = bp^{k-1}x$, and $yx = cp^{k-1}x$ for suitable integers a, b, c. Suppose $c \not\equiv -b \pmod{p}$. Then either $b \not\equiv 0 \pmod{p}$
or $c \not\equiv 0 \pmod{p}$. Let $w = ax - (b + c)y$. Then either $xw \notin \langle w \rangle$ or
$wx \notin \langle w \rangle$. Thus $xy = -yx$. Finally suppose char $y < p^k$. Since char
$N = p^k$ there exists $z \in N$ with char $z = p^k$. Then char $(y + z) = p^k$,
and the above argument gives $x(y + z) = -(y + z)x$ and $xz = -zx$.
$xy = -yx$ follows. This completes the proof.

4.6.17   THEOREM   *Let R be a nilpotent p-ring with R^2 cyclic, $|R^2| > p$.
Then R is an H-ring if and only if R contains an element x and a subgroup
N with $R = \{x\} \dotplus N$, $px^2 = p^k x$ and char $x = p^{k+m}$ for suitable integers
$k > 1$ and $m > 0$, char $N \leq p^k$, and N almost annihilates R with respect
to $p^{k+m-1}x$. If, moreover, char $N = p^k$, then $m = 1$ and char $\mathfrak{A}(R) < p^k$.*

The proof requires the following

4.6.18 LEMMA *Let R be a nilpotent H-p-ring with $pR^2 \neq 0$. Then any basis for R contains an element x with $px^2 = \alpha p^k x$ and char $x = p^{k+m}$ for suitable integers $\alpha \not\equiv 0 \pmod{p}$, $k > 1$, and $m > 0$, and, for N the subgroup of R spanned by all basis members except x, char $N \leq p^k$ and $pNR = pRN = 0$.*

Proof By **4.6.13** products of distinct basis members for R have characteristic at most p. Since $pR^2 \neq 0$, there must then exist some basis member x with $px^2 \neq 0$. By **4.6.12** we obtain $px^2 = \alpha p^k x$ and char $x = p^{k+m}$ for suitable integers $\alpha \not\equiv 0 \pmod p$, $k > 1$, and $m > 0$.

Let N be the subgroup of R spanned by all basis members except x. To complete the proof it is sufficient to show, for all $y \in N$, that char $y \leq p^k$ and $py^2 = 0$. Without loss of generality we shall assume that $\alpha = 1$, and that $py^2 = p^r y$ and char $y = p^{r+s}$ for some integers $r > 0$, $s \geq 0$. If $r \geq k$ then $x(px + py) \in \langle px + py \rangle$ implies $r = k$ and $s = 0$, so the conclusion holds. If $r < k$ then $y(px + py) \in \langle px + py \rangle$ implies $s = 0$, so the conclusion holds. This completes the proof.

Proof of **4.6.17** It is easy to check that a nilpotent p-ring R satisfying the conclusion of **4.6.17** is an H-ring. For the converse let R denote a nilpotent H-p-ring with R^2 cyclic, $|R^2| > p$. Choose an element $x \in R$ and a subgroup $N \subseteq R$ which satisfy **4.6.18** with $\alpha = 1$. Since $pNR = pRN = 0$ and R^2 is cyclic, it follows that
$$NR + RN \subseteq \{p^{k+m-1}x\}.$$
Since $N \cap \{p^{k+m-1}x\} = 0$, N almost annihilates R with respect to $p^{k+m-1}x$.

Now suppose that char $y = p^k$ for some $y \in N$. Then
$$x(px + y) \in \langle px + y \rangle$$
implies $m = 1$ and $y \notin \mathfrak{A}(R)$. This completes the proof.

We conclude this section with some properties of nilpotent H-p-rings R for which R^2 is not cyclic. The first is an immediate consequence of **4.6.11** and **4.6.18**. For the following results R will denote a nilpotent H-p-ring for which R^2 is not cyclic.

4.6.19 COROLLARY *R^2 has type $_p(n, 1)$ for some integer $n \geq 1$.*

4.6.20 THEOREM *$R = S \dotplus N$ for some additive subgroups S and N such that $R^2 \subseteq S$, S has rank 2, and N almost annihilates R with respect to an element of S.*

Proof First we show that R contains additive subgroups S and N with $R = S \dotplus N$, $R^2 \subseteq S$, and rank $S = 2$. If R^2 has type $_p(1, 1)$ this follows easily from the structure theory for finite abelian p-groups, so suppose $pR^2 \neq 0$. By **4.6.18** we obtain $R = \{x\} \dotplus M$ for some $x \in R$ with $px^2 = p^k x \neq 0$ and some additive subgroup $M \subseteq R$ with $pMR = pRM = 0$ and char $M \leq p^k$.

Suppose x is of type I. Since R^2 is not cyclic there exists $z \in M$ with $0 \neq z \in R^2$, and, since char $z = p$, there is a direct sum decomposition $M = \{y\} \dotplus N$ with $z \in \{y\}$. For $S = \{x, y\}$ we obtain $R = S \dotplus N$, $R^2 \subseteq S$, and rank $S = 2$.

Suppose that x is of type II. Since char $(x^2 - p^{k-1}x) = p$ we obtain $R = \{x\} \dotplus \{y\} \dotplus N$ for some $y \in R$ and subgroup $N \subseteq R$, with $\langle x \rangle \subseteq \{x, y\}$. Since char $\{y, N\} \leq p^k <$ char x, **4.6.15** implies that $y^2 \in \langle x \rangle$, and $z^2 \in \langle x \rangle$ for all $z \in N$. $R^2 \subseteq \langle x \rangle$ follows. Thus, for $S = \{x, y\}$, we again obtain $R = S \dotplus N$, $R^2 \subseteq S$, and rank $S = 2$.

Finally, to show that N almost annihilates R, observe that $pNR = pRN = 0$, and that, for $z \in N$, $zR + Rz \subseteq \langle z \rangle \cap R^2 = \{z^2\}$ since $N \cap R^2 = 0$. Thus the proof is complete.

Combining **4.6.20** with **4.6.10** we obtain:

4.6.21 COROLLARY $R = T \oplus N$ for subrings T and N such that T has rank at most 4 and $N \subseteq \mathfrak{A}(R)$.

The classification of all H-p-rings is now reduced to the enumeration of the possible defining relations for the nilpotent H-p-rings T of rank at most 4. This enumeration is relatively straightforward, although lengthy. For the details see Kruse [1, 3].

Historical remark The study of H-rings was begun in 1945 by M. Šperling [1], who gave an incomplete list of the H-rings generated by one element. This list was corrected by P. A. Freĭdman [1] and independently found by L. Rédei [2]. Complete characterizations of H-rings in terms of generators and relations have been independently found by Kruse [1, 2, 3] and by V. I. Andrijanov [1, 2], Andrijanov and Freĭdman [1]. The present treatment follows Kruse [2, 3]. The related class of rings with the property that every proper subring is an ideal in a properly larger subring has been studied by Freĭdman [2, 3, 4]. Rings in which every additive subgroup is an ideal have been characterized by Rédei [1], and by A. Jones and J. J. Schäffer [1].

CHAPTER V

Counting finite nilpotent rings

1 Introduction

THE MAIN OBJECT of this chapter is to obtain some order-of-magnitude bounds on the number of nilpotent rings of order p^n as a function of p and n. It is easy to show that this number increases very rapidly with n. There are, for example, more than 100,000 mutually non-isomorphic nilpotent rings of order 64. To prove this, observe that the proof of Theorem **5.2.1**, below, yields more than 35,000 non-isomorphic 2-algebras A with dim $A = 6$, dim $A^2 = 2$, $A^3 = 0$. To each of these corresponds, by the family construction in Theorem **3.2.10**, at least one ring of type $_2(2, 1, 1, 1, 1)$ and of type $_2(2, 2, 1, 1)$.

The principal result of this chapter is obtained in §2. We shall show that, if the number of nilpotent rings of order p^n is written in the form $p^{f(n,p)n^3}$, then $\lim_{n\to\infty} f(n, p) = 4/27$ independent of p. First we obtain a lower bound for $f(n, p)$ by studying finite nilpotent algebras of exponent 3. Second we reduce the problem of finding an upper bound on $f(n, p)$ to the analogous problem for algebras. Third we find the upper bound for algebras by studying certain properties of algebras of exponent 3.

In §3 we obtain asymptotic bounds on the number of non-nilpotent rings of order p^n. Our purpose in studying non-nilpotent rings is two-fold. First, we wish to illustrate how a result for general rings may be obtained by combining results for radical rings and for semi-simple rings, and then studying the extension problem, the construction of rings with a given radical and given semi-simple factor ring. Second, we wish to show that "almost all" finite rings are "nearly" nilpotent, in the following sense. Let $f(n, p)$ denote the number of pairwise non-isomorphic rings of order p^n, and, for $\varepsilon > 0$, let $g(n, p, \varepsilon)$ denote the number of pairwise non-isomorphic rings R of order p^n such that the order of R modulo its radical is at least $p^{n\varepsilon}$. Then, for every prime p and every $\varepsilon > 0$,

$$\lim_{n\to\infty} \frac{g(n, p, \varepsilon)}{f(n, p)} = 0.$$

In this chapter we shall use the notation $f(n) = 0(n^\alpha)$ to mean that there exists a real number C such that $|f(n)| \leq Cn^\alpha$ for all natural numbers n. The bound C is independent of any other parameters on which the function $f(n)$ may depend.

Throughout this chapter A will denote a nilpotent q-algebra of dimension n, where q is a prime-power. A fundamental tool used in this chapter is the representation of A as a subalgebra of the strictly lower triangular $(n + 1) \times (n + 1)$ matrices over $GF(q)$. This representation was described in Theorem **2.1.2**, and requires, for $e = \exp A$, that the corresponding basis for A be chosen by first choosing a basis for A^{e-1}, extending this to a basis for A^{e-2}, and so on. Further, for $t = \dim A^2$, the algebra A is clearly determined up to isomorphism by specifying the entries in the $n - t$ matrices corresponding to the $n - t$ basis members for A which are not in A^2.

2 Asymptotic results

To obtain a lower bound for the number of nilpotent rings of order p^n we shall first study nilpotent algebras of exponent 3. The following result is analogous to a theorem of G. Higman [1] for p-groups.

5.2.1 THEOREM *Let q be a prime-power and let $f(n, r, q)$ denote the number of pairwise non-isomorphic q-algebras A such that* $\dim A = n$, $\dim (A/A^2) = r$, *and* $A^3 = 0$. *If* $r^2 < n - r$ *then* $f(n, r, q) = 0$. *If* $r^2 \geq n - r$ *then*
$$q^{r^2(n-r) - (n-r)^2 - r^2} \leq f(n, r, q) \leq q^{r^2(n-r) - (n-r)^2 + n - r}.$$

Proof Let $\mathscr{A}$ be a set of pairwise non-isomorphic q-algebras A with $\dim A = n$, $\dim (A/A^2) = r$, and $A^3 = 0$. Consider $A \in \mathscr{A}$ as a homomorphic image of the free nilpotent q-algebra of exponent 3 with r generators (see §2.2), which we denote by F. Since $\dim F^2 = r^2$, $\dim A^2 = n - r \leq r^2$. Thus $f(n, r, q) = 0$ if $r^2 < n - r$. Suppose $r^2 \geq n - r$. Each algebra $A \in \mathscr{A}$ is isomorphic to F/I for some ideal I of F, $I \subseteq F^2$, $\dim I = r^2 - (n - r)$. Hence the number of algebras in $\mathscr{A}$ is bounded above by the number of subalgebras of F^2 of dimension $r^2 - (n - r)$. Since F^2 is a vector space over $GF(q)$, duality implies that this number is the same as the number of subalgebras of dimension $n - r$. This number is
$$\frac{(q^{r^2} - 1)(q^{r^2} - q) \cdots (q^{r^2} - q^{n-r-1})}{(q^{n-r} - 1)(q^{n-r} - q) \cdots (q^{n-r} - q^{n-r-1})},$$
which is less than $q^{r^2(n-r) - (n-r-1)(n-r)}$. Thus the right hand inequality of the Theorem holds.

To obtain a lower bound for $f(n, r, q)$ let us select an algebra $A \in \mathcal{A}$ and a basis $\{e_1, \ldots, e_n\}$ for A such that $A^2 = \langle e_1, \ldots, e_{n-r} \rangle$. Then, for suitable structure constants $\varphi_{ijk} \in GF(q)$, $1 \leq k \leq n - r$, $n - r + 1 \leq i, j \leq n$,

$$e_i e_j = \sum_{k=1}^{n-r} \varphi_{ijk} \, e_k.$$

Moreover, A is determined up to isomorphism by the choice of the φ_{ijk}, and the number of possible choices is $q^{r^2(n-r)}$. Two different bases for A, however, may lead to different determinations of the φ_{ijk}. The number of bases for A^2 is clearly $| \, GL(n - r, q) \, |$, where, for any $m \geq 1$, $GL(m, q)$ denotes the group of all non-singular $m \times m$ matrices over $GF(q)$. Each basis for A^2 may be completed to a basis for A in $| \, GL(r, q) \, | \, q^{r(n-r)}$ ways. Two completions, however, whose corresponding basis elements differ only by an element of A^2, must define the same structure constants φ_{ijk}. Thus we obtain the inequality

$$f(n, r, q) \geq \frac{q^{r^2(n-r)}}{| \, GL(n - r, q) \, || \, GL(r, q) \, |}.$$

However, for $m \geq 1$ we have

$$| \, GL(m, q) \, | = (q^m - 1) \, (q^m - q) \, \ldots \, (q^m - q^{m-1}) \leq q^{m^2}.$$

Hence $f(n, r, q) \geq q^{r^2(n-r) \, - \, (n-r)^2 - r^2}$, so the left hand inequality holds.

5.2.2 THEOREM *If the number $f(n, q)$ of pairwise non-isomorphic nilpotent q-algebras of exponent 3 and dimension $n \geq 2$ is written in the form q^α, then $\alpha = 4n^3/27 + 0(n^2)$.*

Proof Let r be the greatest integer not exceeding $2n/3$. By Theorem **5.2.1** the number of pairwise non-isomorphic q-algebras A with $\dim A = n$, $\dim (A/A^2) = r$, and $A^3 = 0$ is at least

$$q^{r^2(n-r)-(n-r)^2-r^2} = q^{4n^3/27+0(n^2)}.$$

$\alpha \geq 4n^3/27 + 0(n^2)$ follows directly.

For the reverse inequality, we have, from Theorem **5.2.1**,

$$f(n, q) = \sum_{r=1}^{n-1} f(n, r, q) \leq (n - 1)q^{\mu+0(n^2)},$$

where $\mu = \max \, \{r^2(n - r) \mid 1 \leq r \leq n - 1\}$. To find μ let us for the moment allow r to vary continuously over the interval $[0, n]$. Then, for $g(r) = r^2(n - r)$, we have the derivative $g'(r) = 2r(n - r) - r^2$. Thus $g'(r) = 0$ implies $r = 0$ or $r = 2n/3$. Since $g(0) = g(n) = 0$, while $g(2n/3) = 4n^3/27$, g takes on this maximum at $r = 2n/3$. Since, finally, $n - 1 = 0(q^n)$, we have $\alpha \leq 4n^3/27 + 0(n^2)$. Thus the proof is complete.

5.2.3 COROLLARY *The number of pairwise non-isomorphic nilpotent rings of order p^n is at least p^α, where $\alpha = 4n^3/27 + 0(n^2)$.*

The rest of this section is devoted to showing that the number of pairwise non-isomorphic nilpotent rings of order p^n is at most p^α, $\alpha - 4n^3/27 + 0(n^{8/3})$. We shall first show that it is sufficient to consider only the case of algebras.

5.2.4 THEOREM *Suppose the number of pairwise non-isomorphic nilpotent p-algebras of dimension n is written in the form p^α. Then the number of pairwise non-isomorphic nilpotent rings of order p^n is less than $p^{\alpha + n^2 + n}$.*

Proof Let R be a nilpotent ring of order p^n. Let $\{x_1, \ldots, x_m\}$ be a basis for R, with char $x_i = p^{k_i}$, $1 \leq i \leq m$, so that $\sum_{i=1}^m k_i = n$. Next, for $1 \leq i \leq m$ and $0 \leq j < k_i$, define $y_{ij} = p^j x_i$. Rename the y_{ij} in any convenient (predetermined) ordering as $z_1, \ldots, z_n$. Then there are integers φ_{ijk}, $1 \leq i, j, k \leq n$, $0 \leq \varphi_{ijk} < p$, such that

$$z_i z_j = \sum_{k=1}^n \varphi_{ijk} z_k.$$

Moreover, the ring R is determined up to isomorphism by the structure constants φ_{ijk}.

Now let us define a p-algebra A with a basis $\{e_1, \ldots, e_n\}$ by setting

$$e_i e_j = \sum_{k=1}^n \varphi_{ijk} e_k.$$

By construction the multiplicative semigroups of A and R are isomorphic, so associativity and nilpotence of A are equivalent to associativity and nilpotence of R. Hence for any nilpotent ring of order p^n we can construct a nilpotent p-algebra of dimension n and an associated basis for the algebra. Reversing the above process, for any nilpotent p-algebra of dimension n with a given basis we can construct at most one nilpotent ring of type $_p(k_1, \ldots, k_m)$. (In general, the "ring" R constructed may fail to satisfy the distributive law.) The number of nilpotent p-algebras of dimension n is by hypothesis p^α; the number of distinct bases of each algebra is $| GL(n, p) | \leq p^{n^2}$. Hence the number of nilpotent rings of type $_p(k_1, \ldots, k_m)$ is at most $p^{\alpha + n^2}$. The number of abelian groups of order p^n, finally, equals the number of partitions of n, which is less than $2^{n-1} < p^n$. Thus the number of nilpotent rings of order p^n is less than $p^{\alpha + n^2 + n}$.

5.2.5 CorollARY *If the number of all pairwise non-isomorphic p-algebras of dimension n is p^α, then the number of all pairwise non-isomorphic rings of order p^n is less than $p^{\alpha+n^2+n}$.*

Proof Delete all occurrences of the word "nilpotent" from the statement and proof of **5.2.4**.

For the remainder of this section we shall consider only algebras. Next we shall use the matrix representation in a very simple way to obtain an upper bound on the number of pairwise non-isomorphic nilpotent q-algebras of dimension n.

5.2.6 *The number of pairwise non-isomorphic nilpotent q-algebras of dimension n is at most q^α, $\alpha = (n^3/3\sqrt{3}) + 0(n)$.*

Remark The bound of $1/(3\sqrt{3})$ is about 0.192, compared to the lower bound of $4/27 \approx 0.148$.

Proof Let A be a nilpotent q-algebra of dimension n. Let $\dim(A/A^2) = r$. In the representation of A in strictly lower triangular matrices (see **2.1.2**), the r generators of A have the form

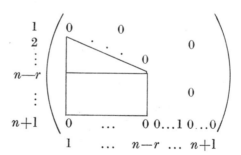

The number of possibly non-zero entries in such a matrix is
$$r(n-r) + \tfrac{1}{2}(n-r-1)(n-r) \leq \tfrac{1}{2}(n^2-r^2).$$
The number of generators is r, and hence the total number of pairwise non-isomorphic algebras A is at the most $q^{f(n,r)}$, where
$$f(n,r) = \tfrac{1}{2}r(n^2-r^2).$$
Let us for the moment allow r to vary continuously on the interval $[0, n]$. Then
$$\frac{\partial f}{\partial r} = \tfrac{1}{2}(n^2-r^2) - r^2$$
Thus the only interior critical point occurs when $r = n/\sqrt{3}$. Since

$f(n, 0) = f(n, n) = 0$ and $f(n, n/\sqrt{3}) = n^3/3\sqrt{3}$, $f(n, r) \leq n^3/3\sqrt{3}$ for all r, $r = 0, 1, \ldots, n$. The number of choices for $\dot{r}$ is $n + 1 = q^{0(n)}$. Hence the total number of pairwise non-isomorphic nilpotent q-algebras of dimension n is at most q^α, $\alpha = (n^3/3\sqrt{3}) + 0(n)$.

In order to obtain the best possible bound on the number of pairwise non-isomorphic nilpotent q-algebras A of dimension n we shall study some properties of subalgebras B minimal subject to $B^2 = A^2$. This idea and the key steps of its development are due to C. Sims [1], who studied the analogous problem of obtaining asymptotic bounds on the number of groups of order p^n, proving that this number is p^α, $\alpha = 2n^3/27 + 0(n^{8/3})$.

5.2.7 THEOREM *Let A be a finite-dimensional nilpotent algebra such that no proper subalgebra B satisfies $B^2 = A^2$. Then*
$$\dim (A/A^2) \leq \dim (A^2/A^3) + 1.$$

Proof Without loss of generality we assume $A^3 = 0$. Let $r = \dim (A/A^2)$. It is clearly sufficient for us to construct a set of elements of A, $\{x_1, \ldots, x_r\}$, which are linearly independent mod A^2 and such that the sequence of subalgebras of A^2
$$\langle x_1\rangle^2 \subset \langle x_1, x_2\rangle^2 \subset \ldots \subset \langle x_1, \ldots, x_r\rangle^2$$
is properly ascending.

To construct the desired set we choose for x_1 any element of A not in A^2 and proceed by induction. Suppose, for some $k \geq 1$, that $x_1, \ldots, x_k$ have been chosen so that $\langle x_1, \ldots, x_i\rangle^2 \neq \langle x_1, \ldots, x_{i+1}\rangle^2$ for any i, $1 \leq i < k$. If $k = r$ we are done. If not, then, by the Burnside basis theorem **4.1.4**, $B = \langle x_1, \ldots, x_k\rangle$ is a proper subalgebra of A, so by hypothesis B^2 is proper in A^2. If there exists an element $y \in A$ such that $y^2 \notin B^2$ or such that $yB + By \nsubseteq B^2$, then we shall define $x_{k+1} = y$. In the contrary case there exist, since $B^2 \neq A^2$, elements y, $z \in A - B$, with $yz \notin B^2$, but y^2, yB, and By contained in B^2. If $k = 1$ we replace x_1 by $x_1' = y$ and define $x_2 = z$. Assume $k > 1$ and define $C = \langle x_1, \ldots, x_{k-1}\rangle$. If $\langle C, y\rangle^2 \neq C^2$ then replace x_k by $x_k' = y$ and define $x_{k+1} = z$. If $\langle C, y\rangle^2 = C^2$ then replace x_k by $x_k' = x_k + y$ and define $x_{k+1} = z$. In each of these cases the elements $x_1, \ldots, x_{k-1}, x_k', x_{k+1}$ are linearly independent mod A^2, and the inclusions $C^2 \subset \langle C, x_k'\rangle^2 \subset \langle C, x_k', x_{k+1}\rangle^2$ are proper. This completes the proof.

5.2.8 COROLLARY *Let A be a nilpotent algebra, and let $t = \dim (A^2/A^3)$. Then A contains a subalgebra B such that $B^2 = A^2$ and $\dim (B/A^2) \leq t + 1$.*

Proof   By **5.2.7** A contains a subalgebra B such that $B^2 + A^3 = A^2$ and dim $(B/B^2) \leq t + 1$. By **1.3.3** $B^2 = A^2$.

To show that the bound in Theorem **5.2.7** cannot be improved, consider the following example. Choose an integer $t > 1$ and let A be an algebra with a basis $\{x_1, \ldots, x_{t+1}, y_1 \ldots, y_t\}$ such that $y_i \in \mathfrak{A}(A)$, $1 \leq i \leq t$, $x_1^2 = 0$, $x_1 x_i = -x_i x_1 = y_{i-1}$ for $2 \leq i \leq t + 1$, and $x_i x_j = 0$ if $i \geq 2$ and $j \geq 2$. Then dim $(A/A^2) = t + 1 = \dim A^2 + 1$. To show that no proper subalgebra B of A satisfies $B^2 = A^2$ first define $N = \langle x_2, \ldots, x_{t+1}, A^2 \rangle$. Then either $B \subseteq N$, in which case $B^2 = 0$, or else $B \cap N$ has codimension 1 in B. In the second case choose a basis $\{b_1, \ldots, b_r\}$ for B so that $B \cap N = \langle b_2, \ldots, b_r \rangle$. Then B^2 is generated by the elements $b_1 b_2, \ldots, b_1 b_r$, and so has dimension at most $r - 1 \leq t - 1$. Hence $B^2 \neq A^2$.

To obtain an upper bound for the number of q-algebras A of dimension n we could at this point count entries in matrices in a way similar to the proof of **5.2.6**, where, in choosing the basis elements of A which are not in A^2, we first complete a basis for a subalgebra B of minimum dimension such that $B^2 = A^2$ and then complete the basis for A. Using **5.2.8** we can then obtain an upper bound of the form $q^{\alpha n^3 + 0(n^2)}$, where $\alpha \approx 0.157$, compared to the lower bound of $\alpha = 4/27 \approx 0.148$. To obtain the best possible upper bound we need one further result.

5.2.9   THEOREM   *Let $\mathscr{A}$ be a set of pairwise non-isomorphic q-algebras A such that dim $(A/A^2) = r$, dim $A^2 = t$, $A^3 = 0$, and s is the minimum dimension of B/A^2 as B ranges over the subalgebras of A such that $B^2 = A^2$. Then the number of algebras in $\mathscr{A}$ is no more than q^α, where $\alpha = r^2(t - s) + 0((r + t)^{8/3})$.*

The proof requires the following

5.2.10   LEMMA   *Let $A \in \mathscr{A}$, r, s, and t be as defined above. Then, for every subalgebra B of A such that $B \supseteq A^2$,*
$$\dim B^2 - \dim (B/A^2) \leq t - s + 1.$$

Proof   Suppose the lemma is false, and let $B \supseteq A^2$ be a maximal subalgebra of A subject to
$$\dim B^2 - \dim (B/A^2) \geq t - s + 2. \tag{1}$$
If $B^2 = A^2$ then dim $B^2 = t$, so (1) implies dim $(B/A^2) \leq s - 2$, contradicting the definition of s. Thus $B^2 \neq A^2$. Choose a subalgebra C of A such that $B + C = A$ and $B \cap C = A^2$. We now show that $BC + CB \subseteq B^2$. For suppose there exists $c \in C$ with $Bc + cB \nsubseteq B^2$. Then

$\dim \langle B, c\rangle^2 - \dim (\langle B, c\rangle/A^2) \geq (\dim B^2 + 1)$
$$- (\dim (B/A^2) + 1) \geq t - s + 2,$$
contradicting the maximality of B. Thus $BC + CB \subseteq B^2$, so $A^2 = (B + C)^2 = B^2 + C^2$. Thus the projection of C^2 into A^2/B^2 is onto, and hence, for $\bar{C} = C/B^2$, $\dim \bar{C}^2 = \dim A^2 - \dim B^2$. Applying Corollary **5.2.8** to $\bar{C}$ we obtain a subalgebra $\bar{D}$ of $\bar{C}$ such that $\bar{D}^2 = \bar{C}^2$ and $\dim (\bar{D}/\bar{D}^2) \leq \dim \bar{C}^2 + 1$. Let D be the pre-image of $\bar{D}$ in A. Then $D^2 = C^2$, so $D \supseteq B^2 + D^2 = B^2 + C^2 = A^2$, and

$\dim (D/A^2) \leq \dim \bar{C}^2 + 1 = \dim A^2 - \dim B^2 + 1 =$
$$t - \dim B^2 + 1. \qquad (2)$$

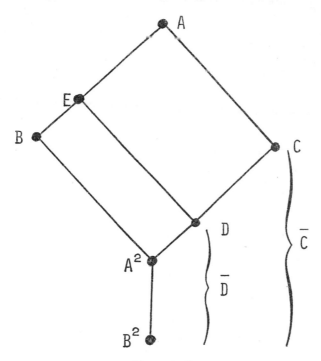

FIGURE 6

Finally define $E = B + D$. Then $E^2 = B^2 + D^2 = B^2 + C^2 = A^2$ and, by (1) and (2),

$\dim (E/A^2) = \dim (B/A^2) + \dim (D/A^2)$
$\leq (\dim B^2 - t + s - 2) + (t - \dim B^2 + 1)$
$= s - 1,$

which contradicts the definition of s. Thus the lemma is proved.

Proof of Theorem   For an algebra $A \in \mathscr{A}$ choose elements $x_1, \ldots, x_r$ in A which map onto a basis for A/A^2. The algebra A is determined up to isomorphism by the products $x_i x_j$, $1 \leq i, j \leq r$. We shall count the number of ways of choosing these products.

Let u be the greatest integer not exceeding $r^{2/3}$, and let v be the smallest integer not less than r/u. For $1 \leq i \leq v - 1$ define
$$A_i = \{x_{(i-1)u+1}, \ldots, x_{iu}\}$$
and for $i = v$ define
$$A_i = \{x_{(i-1)u+1}, \ldots, x_r\}.$$
Let us first observe by the lemma that dim $\langle A_i \cup A_j \rangle^2 \leq w$, where $w = \min \{t, t - s + 1 + 2u\}$, for $1 \leq i < j \leq v$. Next let us choose the w-dimensional subspaces B_{ij} of A^2 which are, respectively, to contain the subspaces $\langle A_i \cup A_j \rangle^2$, $1 \leq i < j \leq v$. The number of w-dimensional subspaces of A^2 is

$$\frac{(q^t - 1)(q^t - q) \cdots (q^t - q^{w-1})}{(q^w - 1)(q^w - q) \cdots (q^w - q^{w-1})} \leq q^{t^2}.$$

The number of ways of choosing i and j, $1 \leq i < j \leq v$, is $\frac{1}{2}v(v-1)$. Hence the number of ways of choosing all of the spaces B_{ij} together is at the most $q^{(1/2)t^2 v(v-1)} = q^{0((r+t)^{8/3})}$.

Once the subspaces B_{ij} have been chosen, the number of ways of choosing each product $x_k x_l$, $1 \leq k, l \leq r$, is q^w, since $x_k x_l$ must lie in a predetermined subspace B_{ij}. Hence the total number of algebras $A \in \mathscr{A}$ is no more than q^α, where
$$\alpha = wr^2 + 0((r+t)^{8/3}) = r^2(t - s) + 0((r+t)^{8/3}).$$

We can now prove the main result of this section.

5.2.11 **THEOREM** *If the number of pairwise non-isomorphic nilpotent q-algebras of dimension n is written in the form $q^{\alpha n^3}$, then $\alpha = 4/27 + 0(n^{-1/3})$.*

Proof   That $q^{\alpha n^3}$ is a lower bound follows from **5.2.2**. To show $q^{\alpha n^3}$ is an upper bound, let A be a nilpotent q-algebra of dimension n. Let $r = \dim (A/A^2)$, $t = \dim (A^2/A^3)$, and $u = \dim A^3$, so that $u = n - r - t$. Let s be the minimum dimension of B/A^2 as B ranges over the subalgebras of A such that $B^2 = A^2$. Construct a basis $\{e_1, \ldots, e_n\}$ for A by choosing a basis for A^{e-1}, where $e = \exp A$, extending this to a basis for A^{e-2}, and continuing in this way until a basis for A^2 has been constructed. Next extend the basis to a subalgebra B of A such that $B^2 = A^2$ and dim $(B/A^2) = s$. Finally complete the basis for A. In the corresponding matrix representation the r generators e_i, $n - r + 1 \leq i \leq n$, have the form

Let us first apply Theorem **5.2.9** to the algebra A/A^3. Doing so we find that the number of choices for all of the blocks D_i together is at the most q^{α}, $\alpha = r^2(t - s) + 0(n^{8/3})$.

Second let us select basis elements e_i and e_j, $u + t + s + 1 \leq i \leq n$, $1 \leq j \leq u + t$. Thus $e_i \notin B$ while $e_j \in A^2$. Since $B^2 = A^2$, the values of $e_i e_j$ and $e_j e_i$ are determined by associativity when we are given the values of $e_i e_k$ and $e_k e_i$ for all generators e_k of B,

$$u + t + 1 \leq k \leq u + t + s.$$

Hence the blocks A_i and B_i are determined in terms of the other blocks and matrices when $u + t + s + 1 \leq i \leq n$.

The number of choices for all of the r blocks C_i of the r generators of A is clearly $q^{r^2 u}$. The number of choices for the blocks A_i and B_i for the s generators of B is $q^{su(t+\frac{1}{2}(u-1))}$. Multiplying these terms together we obtain that the number of algebras A is bounded above by q^{α}, where

$$\alpha = r^2(t - s) + r^2 u + su(t + \tfrac{1}{2}u) + 0(n^{8/3}).$$

Recall that $u = n - r - t$, and define the function

$$f(r, s, t) = r^2(t - s) + r^2(n - r - t) + s(n - r - t)(t + \tfrac{1}{2}(n - r - t))$$
$$= r^2(n - r - s) + \tfrac{1}{2}s((n - r)^2 - t^2).$$

Since $\dfrac{\partial f}{\partial t} = -st < 0$, for fixed r and $s, f(r, s, t)$ takes on its maximum

for the smallest possible t, which by **5.2.8** is $t = s - 1$. Since the quantity $\tfrac{1}{2}s(t^2 - (t - 1)^2) = 0(n^2)$ it is sufficient for us to find the maximum of $f(r, s, s)$. The requirement that no dimension be negative implies that r and s satisfy the inequalities

$$0 \leq s \leq r \leq n \quad \text{and} \quad r + s \leq n.$$

G

Further, $f(r, s, s) = r^2(n - r - s) + \frac{1}{2}s((n - r)^2 - s^2)$
$$= (n - r - s)(r^2 + \frac{1}{2}s(n - r + s))$$
$$\leq (n - r - s)(r^2 + \frac{1}{2}ns)   \text{since } s \leq r.$$

Define $g(r, s) = (n - r - s)(r^2 + \frac{1}{2}ns)$. Then

$$\frac{\partial g}{\partial s} = -(r^2 + \frac{1}{2}ns) + \frac{1}{2}n(n - r - s)$$

so at a critical point of g we have $s = (n^2 - nr - 2r^2)/2n$.

Moreover, $\dfrac{\partial g}{\partial r} = 2r(n - r - s) - (r^2 + \frac{1}{2}ns)$

so at a critical point of g we have $s = 2r(2n - 3r)/(n + 4r)$. Hence at a critical point we have

$$(n^2 - nr - 2r^2)(n + 4r) = 2n(2r)(2n - 3r),$$

or $(n - 4r)(n^2 - nr + 2r^2) = 0.$

Since $n^2 - nr + 2r^2$ has no real root r, the only possible critical point occurs for $r = n/4$. This, however, gives a value of $s = (5/4)r$, which is not within the prescribed bounds. Hence $g(r, s)$ has no critical points, so its maximum occurs on the boundary of its domain. The boundary is the triangle bordered by the lines $s = 0$, $s = r$, and $r + s = n$. For $r + s = n$ we have $g(r, s) = 0$. For $s = r$, $g(r, r) = \frac{1}{2}r(n^2 - 4r^2)$, which takes on its maximum of $n^3/6\sqrt{3} < 4n^3/27$ for $r = n/2\sqrt{3}$. Finally, for $s = 0$, $g(r, 0) = r^2(n - r)$, which takes on its maximum of $4n^3/27$ for $r = 2n/3$.

We have now shown that for each possible choice of r, s, and t the number of algebras A is at the most q^α, where $\alpha = 4n^3/27 + 0(n^{8/3})$. Finally, let us note that the number of choices for r, s, and t is less than $n^3 = 0(q^n)$. Hence the proof is complete.

Combining **5.2.4** with **5.2.11** we obtain:

5.2.12   COROLLARY   *If the number of pairwise non-isomorphic nilpotent rings of order p^n is written in the form $p^{\alpha n^3}$, then $\alpha = 4/27 + 0(n^{-1/3})$.*

3 Asymptotic results for non-nilpotent rings

To obtain the results of this section it will suffice, by **5.2.5**, for us to study only finite algebras. The fundamental structure theory developed by Wedderburn for finite-dimensional semi-simple algebras will be of constant use. For convenience of reference we quote Wedderburn's results.

5.3.1 THEOREM *Let A be a finite-dimensional algebra and let N be the radical of A. Then A contains a subalgebra B such that $B \cong A/N$. (It follows that B is semi-simple, $B + N = A$, and $B \cap N = 0$.)*

5.3.2 THEOREM *A finite-dimensional semi-simple algebra is uniquely decomposable as a direct sum of simple subalgebras.*

5.3.3 THEOREM *A finite-dimensional simple algebra over a field F is isomorphic to a complete n^2-dimensional algebra of matrices over a division ring which contains F in its center.*

5.3.4 THEOREM *Every finite division ring is commutative.*

Theorems **5.3.1**—**5.3.3** are from Wedderburn [2], while **5.3.4** is from Wedderburn [1]. Modern proofs of **5.3.2**—**5.3.4** may be found, for example, in Herstein [3] or in van der Waerden [1]. A proof of **5.3.1** may be found in Albert [1] or in Jacobson [1].

5.3.5 THEOREM *The number of pairwise non-isomorphic semi-simple q-algebras of dimension $m \geq 1$ is less than $2^{m-1} e^{m/e}$, where $e = 2.7182\ldots$ is the base for natural logarithms.*

Proof The number of ways of choosing the number of simple components of a semi-simple q-algebra of dimension m, together with their dimensions over $GF(q)$, is the number of unordered partitions of m, which does not exceed the number of ordered partitions, 2^{m-1}. The number of pairwise non-isomorphic simple q-algebras of dimension r is the number of factorizations $r = s^2 l$, where s and t are natural numbers. This number is at most r. Thus the number of pairwise non-isomorphic semi-simple algebras corresponding to a given partition
$$m = m_1 + m_2 + \ldots + m_k$$
is at the most $m_1 m_2 \ldots m_k$, which does not exceed $(m/k)^k < e^{m/e}$. The result follows.

5.3.6 THEOREM *Let S be a finite semi-simple q-algebra, and let N be a nilpotent q-algebra of dimension n. Then the number of pairwise non-isomorphic q-algebras A with radical N and $A/N \cong S$ is less than q^{2n^2}.*

Remark Note that the bound does not depend on the dimension of the semi-simple algebra S. Several of the ideas used in the proof of **5.3.6** are similar to those employed by M. Hall [1] in studying a finite-dimensional algebra "bound to its radical". If A is a finite-dimensional algebra with radical N, Hall shows that A decomposes as a direct sum of subalgebras, $A = B \oplus C$, where B is semi-simple, $N \subseteq C$, and dim $C \leq n^2 + n + 1$, where $n = \dim N$.

Proof By **5.3.1** A contains a semi-simple subalgebra B with $B + N = A$, $B \cap N = 0$, and $B \cong S$. Thus A is determined up to isomorphism by the action on N induced by multiplication on the left and on the right by B. To count the number of different actions of B on the right of N, we shall consider N as a module for the semi-simple algebra B and decompose N as a direct sum of submodules for B. The symbol $\dotplus$ will denote a module direct sum.

By **5.3.2** B is a direct sum of ideals, $B = B_1 \oplus \ldots \oplus B_l$, where each B_i is a simple q-algebra, $1 \leq i \leq l$. Define $N_i = NB_i$, $1 \leq i \leq l$, and $N_0 = \{x \in N \mid xB = 0\}$. We now show that

$$N = N_0 \dotplus N_1 \dotplus \ldots \dotplus N_l. \tag{1}$$

For, if we denote by e_i the identity of B_i, $1 \leq i \leq l$, and $e = e_1 + \ldots + e_l$, then we have, for all $x \in N$,

$$x = (x - xe) + xe_1 + xe_2 + \ldots + xe_l.$$

But $(x - xe)B = 0$, so $x - xe \in N_0$, while $xe_i \in N_i$, $1 \leq i \leq l$. Thus N is the sum of the submodules N_i, $0 \leq i \leq l$. Now suppose $x_0 + x_1 + \ldots + x_l = 0$, $x_i \in N_i$, $0 \leq i \leq l$. Then

$$0 = (x_0 + x_1 + \ldots + x_l)e_i = x_i e_i = x_i$$

for $1 \leq i \leq l$, so that $x_1 = \ldots = x_l = 0$. Then also $x_0 = 0$, so the sum of the N_i is direct, and (1) is proved.

Since $e_i e_j = 0$ for $i \neq j$, $N_i B_j = 0$ for $i \neq j$. Thus the action of B on the right of N is determined by the choice of the submodules N_i, $0 \leq i \leq l$, and by the action of B_i on N_i, $1 \leq i \leq l$.

Since B_i is a finite-dimensional simple algebra, every B_i-module is completely reducible, so that, for some integer k_i,

$$N_i = N_{i1} \dotplus N_{i2} \dotplus \ldots \dotplus N_{ik_i},$$

where N_{ij} is an irreducible B_i-module. Moreover, each irreducible B_i-module N_{ij} is B_i-isomorphic to a fixed minimal right ideal R of B_i. Let $f_{ij} : N_{ij} \to R$ be such a B_i-isomorphism. Then, for $x \in N_{ij}$ and $b \in B_i$,

$$xb = f_{ij}^{-1}(f_{ij}(xb)) = f_{ij}^{-1}(f_{ij}(x) \cdot b),$$

so that the action of B_i on the right of N_i is completely determined by the choice of the submodules N_{ij} and the mappings f_{ij}, $1 \leq i \leq l$, $1 \leq j \leq k_i$.

By **5.3.3** and **5.3.4**, each of the simple algebras B_i, $1 \leq i \leq l$, is isomorphic, for some positive integers a_i and n_i, to the algebra of all $n_i \times n_i$ matrices over $F_i = GF(q^{a_i})$. It follows that the minimal right ideal R of B_i has dimension n_i over F_i. Thus dim $_{F_i} N_{ij} = n_i$, so that dim $_{GF(q)} N_{ij} = a_i n_i$ and dim $_{GF(q)} N_i = k_i a_i n_i$ for $1 \leq i \leq l$, $1 \leq j \leq k_i$.

Let $n_0 = $ dim $_{GF(q)} N_0$. Then $n = n_0 + \sum_{i=1}^{l} k_i a_i n_i$.

Next we need the following

5.3.7 LEMMA *Let V be a vector space of dimension n over $GF(q)$, and let $n = n_1 + \ldots + n_l$ be a partition of n into positive parts. Then the number of ordered direct sum decompositions $V = V_1 \dotplus \ldots \dotplus V_l$ into subspaces V_i with* dim $V_i = n_i$, $1 \leq i \leq l$, *is*

$$\frac{\mid GL(n, q) \mid}{\prod\limits_{i=1}^{l} \mid GL(n_i, q) \mid}.$$

Proof For a given ordered direct sum decomposition
$$V = V_1 \dotplus \ldots \dotplus V_l$$
with dim $V_i = n_i$, $1 \leq i \leq l$, we can construct an ordered basis for V by first selecting an ordered basis for V_1, then an ordered basis for V_2, and so on. Since V_i has $\mid GL(n_i, q) \mid$ distinct ordered bases, a given direct sum decomposition of V corresponds in this way to exactly $\prod\limits_{i=1}^{l} \mid GL(n_i, q) \mid$ ordered bases for V. On the other hand, there are, altogether, exactly $\mid GL(n, q) \mid$ ordered bases for V. The result follows.

Proof of **5.3.6**, *continued* The number of choices in N for the submodules N_i, $0 \leq i \leq l$, does not exceed the number of ways of decomposing N as a vector space over $GF(q)$ as a direct sum of subspaces of the appropriate dimensions. By the Lemma this number is

$$\frac{\mid GL(n, q) \mid}{\mid GL(n_0, q) \mid \prod\limits_{i=1}^{l} \mid GL(k_i a_i n_i, q) \mid} \qquad (2)$$

Similarly, the number of choices for the submodules N_{ij} in N_i does not exceed

$$\frac{\mid GL(k_i n_i, q^{a_i}) \mid}{\mid GL(n_i, q^{a_i}) \mid^{k_i}}, \; 1 \leq i \leq l. \qquad (3)$$

The number of possible B_i-isomorphisms f_{ij} of N_{ij} onto R does not exceed the number of $GF(q^{a_i})$-isomorphisms, which is
$$\mid GL(n_i, q^{a_i}) \mid, \; 1 \leq i \leq l, 1 \leq j \leq k_i. \qquad (4)$$
Multiplying expressions (2)—(4) together with the appropriate multiplicities, we find that the number of different possible actions of B on the right of N does not exceed

$$\frac{\mid GL(n, q) \mid \prod\limits_{i=1}^{l} \mid GL(k_i n_i, q^{a_i}) \mid}{\mid GL(n_0, q) \mid \prod\limits_{i=1}^{l} \mid GL(k_i a_i n_i, q) \mid} \leq \mid GL(n, q) \mid,$$

since $\mid GL(k_i a_i n_i, q) \mid \geq \mid GL(k_i n_i, q^{a_i}) \mid$ and $q^{a_i} - 1 \geq 1$. Finally, $\mid GL(n, q) \mid < q^{n^2}$, so the number of actions of B on the right of N is less than q^{n^2}. Dually, the number of actions of B on the left of N is less than q^{n^2}. Thus the proof of **5.3.6** is complete.

Combining Theorems **5.2.11**, **5.3.5**, and **5.3.6** we now obtain immediately:

5.3.8 THEOREM *If the number of pairwise non-isomorphic q-algebras of dimension n is written in the form $q^{\alpha n^3}$, then $\alpha = 4/27 + O(n^{-1/3})$.*

By **5.2.5** we also obtain:

5.3.9 COROLLARY *If the number of pairwise non-isomorphic rings of order p^n is written in the form $p^{\alpha n^3}$, then $\alpha = 4/27 + O(n^{-1/3})$.*

Finally, we can show that "almost all" finite rings are "nearly" nilpotent, as follows:

5.3.10 THEOREM *Let $f(n, p)$ denote the number of pairwise non-isomorphic rings of order p^n, and, for $\varepsilon > 0$, let $g(n, p, \varepsilon)$ denote the number of pairwise non-isomorphic rings R of order p^n such that $\mid R/N \mid \geq p^{n\varepsilon}$, where N is the radical of R. Then, for all $\varepsilon > 0$ and all primes p,*

$$\lim_{n \to \infty} \frac{g(n, p, \varepsilon)}{f(n, p)} = 0.$$

Proof By **5.3.9**, $f(n, p) = p^{\alpha}$, $\alpha = 4n^3/27 + O(n^{8/3})$. By **5.2.11**, **5.3.5**, and **5.3.6**, $g(n, p, \varepsilon) \leq p^{\beta}$, where

$$\beta = (4/27)(1 - \varepsilon)^3 n^3 + O(n^{8/3}) + O(n^2).$$

Thus $g(n, p, \varepsilon)/f(n, p) \leq p^{\gamma}$, where $\gamma = (4/27)[(1 - \varepsilon)^3 - 1]n^3 + O(n^{8/3})$, and $\lim_{n \to \infty} p^{\gamma} = 0$, so the result follows.

The construction of some special rings

FOR CONVENIENCE of reference this chapter collects from the literature some of the more significant calculations of nilpotent rings by the use of generators and relations. A major goal of the chapter is to obtain some information on the nilpotent rings of order p^4. The cyclic rings are trivial. In §1 the power rings of type $_p(2, 2)$ are classified, as well as all nilpotent rings of type $_p(3, 1)$ as a special case of those of type $_p(n, 1)$, $n > 1$. The non-power nilpotent rings of type $_p(2, 2)$ as well as all those of type $_p(2, 1, 1)$ must, by **3.2.8**, lie in the same family as a nilpotent p-algebra of dimension 4, and all such rings lying in a given family can be constructed from the p-algebra by the process given in **3.2.10**. Hence the bulk of this chapter is devoted to the classification of the nilpotent algebras of dimension 4. These computations will be done over an arbitrary field.

The methods employed in this chapter are entirely elementary.

1 Some nilpotent p-rings of rank 2

Throughout this section Greek letters will denote rational integers used as structure constants.

6.1.1 THEOREM *Let R be a power ring of type $_p(2, 2)$. Then R has a basis $\{a, a^2\}$ for which $a^3 = \alpha pa + \beta pa^2$, and α, β satisfy one of the following conditions. These cases are mutually non-isomorphic.*
(1) $\alpha = 0$, $\beta = 0, 1$.
(2) $p = 2$, $\alpha = 1$, $\beta = 0, 1$.
(3) $p \neq 2$, $\alpha = 1, \mu$; $\beta = 0$, *where μ is a fixed non-square* mod p.

Proof Let $R = \langle a \rangle$. Since $R/\langle pa, pa^2 \rangle$ is a power p-algebra of dimension 2, a and a^2 form a basis for R, and $a^3 \in \langle pa, pa^2 \rangle$. Let $a^3 = \alpha pa + \beta pa^2$. Then $a^4 = \alpha pa^2$. Consider a change of basis given by $a' = \zeta a + \eta a^2$, $\zeta \not\equiv 0 \pmod{p}$. Then
$$a'^3 = (\zeta^2 \alpha)\, pa' + (2\eta\alpha + \zeta\beta)\, pa'^2.$$
Thus the cases $\alpha = 0$ and $\alpha \neq 0$ are disjoint. Further, for $\alpha = 0$, β may

be forced 0 or 1, so (1) holds. If $\alpha \neq 0$ and $p = 2$, then β cannot be changed, so (2) holds. If $\alpha \neq 0$ and $p \neq 2$, then an appropriate choice of ζ makes α either 1 or μ, a fixed non-square mod p, and then $\eta \equiv -\zeta\beta(2\alpha)^{-1} \pmod{p}$ makes $\beta = 0$, so (3) holds.

6.1.2 THEOREM *Let R be a nilpotent power ring of type $_p(n, 1)$, $n \geq 2$. Then R has a basis $\{a, b\}$ with $p^n a = pb = 0$, which satisfies one of the following sets of relations. The rings of these cases are mutually non-isomorphic.*
(1) $a^2 = b$, $a^3 = \alpha p^{n-1}a$, $\alpha = 0, 1, \mu$, where μ is a fixed non-square mod p.
(2) $a^2 = p^i a + b$, $ab = ba = \alpha p^{n-1}a$, $0 \leq \alpha < p$, $1 \leq i < n - 1$.

Proof Let a be a generator of R. Since char $R = p^n$, char $a = p^n$. Choose $b \in R$ such that $\{a, b\}$ is a basis for R. Let $a^2 = \gamma p^i a + \delta b$, with $1 \leq i \leq n$ and $\gamma \not\equiv 0 \pmod{p}$. $\delta = 0$ would imply $|\langle a \rangle| = p^n$, so $\delta \neq 0$. By replacing a by $a' = \gamma^{-1}a$, where $\gamma^{-1}\gamma \equiv 1 \pmod{p^n}$, we may assume that $\gamma = 1$. By replacing b by $b' = \delta b$, we may assume $\delta = 1$. Since R is nilpotent, and char $b = p$, $ab = \alpha p^{n-1}a$ for some α, $0 \leq \alpha < p$. $b^2 = 0$ since $p^{n-1}a \in \mathfrak{A}(R)$.

Suppose $i = n$. Then $a^2 = b$, $a^3 = ab = ba = \alpha p^{n-1}a$, and $a^4 = b^2 = 0$. Let $a' = \zeta a + \eta b$, with $\zeta \not\equiv 0 \pmod{p}$. Then
$$a'^3 = \alpha\zeta^3 p^{n-1}a = \alpha\zeta^2 p^{n-1}a'.$$
Thus α may be chosen to be one of 0, 1, or μ, so (1) holds. Further, we have considered the effect of a general change of basis, so the subcases are mutually non-isomorphic.

If $i = n - 1$ then replacement of b by $b' = p^{n-1}a + b$ takes R to the previous case, $i = n$.

Suppose $1 \leq i < n - 1$, and consider a change of basis given by $a' = \zeta a + \eta b$, $\zeta \not\equiv 0 \pmod{p}$. Then
$$a'^2 = \zeta^2 p^i a + \zeta^2 b + 2\zeta\eta\alpha p^{n-1}a$$
$$= (\zeta + 2\eta\alpha p^{n-i-1}) p^i a' + \zeta^2 b.$$
To preserve the relation $a^2 = p^i a + b$ then requires $b' = \zeta^2 b$ and $\zeta + 2\eta\alpha p^{n-i-1} \equiv 1 \pmod{p^{n-i}}$. Thus $\zeta \equiv 1 \pmod{p}$, so $b' = b$, and then $a'b' = ab = \alpha p^{n-1}a'$, so no change in the structure constant α is possible. Thus (2) holds, and the subcases are mutually non-isomorphic.

6.1.3 THEOREM *Let R be a nilpotent ring of type $_p(n, 1)$, $n \geq 2$, not a power ring. Then R has a basis $\{a, b\}$ such that $p^n a = pb = 0$, and a, b satisfy one of the following sets of relations. The cases described are mutually non-isomorphic. μ denotes a fixed non-square mod p.*

(I) (1) $a^2 = b^2 = 0$, $ab = ba = 0$.

  (2) $a^2 = b^2 = 0$, $ab = p^{n-1}a$, $ab + ba = 0$.

  (3) $a^2 = ab = 0$, $ba = b^2 = 2^{n-1}a$, $p = 2$.

(II) (1) $a^2 = p^i a$, $ab = 0$, $ba = 0$, $b^2 = \alpha p^{n-1}a$, $\alpha = 0$, 1, μ,
    $1 \le i \le n - 1$.

  (2) $a^2 = p^i a$, $ab = 0$, $ba = p^{n-1}a$, $b^2 = \alpha p^{n-1}a$, $0 \le \alpha < p$,
    $1 \le i \le n - 1$.

(III) (1) $a^2 = p^i a$, $ab = p^{n-1}a$, $ba = \beta p^{n-1}a$, $b^2 = 0$, $0 \le \beta < p$,
    $1 \le i \le n - 1$.

  (2) $a^2 = ab = p^{n-1}a$, $ba = b^2 = \alpha p^{n-1}a$, $\alpha = \pm 1$.

Proof Let us first define subsets of R

$$S_1 = \{x \in R \mid \text{char } x = p^n\}$$

and $S_2 = \{x \in R \mid \text{char } x = p, x \notin p^{n-1}R\}$.

Note that every basis for R consists of one element from S_1 and one from S_2, and that for all $a \in S_1$, $b \in S_2$, we have ab, ba, $b^2 \in p^{n-1}R$. Further, either $x^2 = 0$ for all $x \in S_2$ or for no $x \in S_2$. Hence, for $\{a, b\}$ a basis for R, $b \in S_2$, the cases $b^2 = 0$ and $b^2 \ne 0$ always lead to non-isomorphic rings. We shall now consider the three cases I, II, and III separately. These are mutually exclusive, and arise in the following way:

I *For all $x \in S_1$, $x^2 = 0$.*

II *There exists $a \in S_1$ with $a^2 \ne 0$, and $b \in S_2$ with $ab = 0$.*

III *There exists $a \in S_1$ with $a^2 \ne 0$, but for all $x \in S_1$ with $x^2 \ne 0$ and all $y \in S_2$, $xy \ne 0$.*

Case I Choose $a \in S_1$ and $b \in S_2$. Then $(a + \zeta b)^2 = 0$ for all integers ζ. For $\zeta \not\equiv 0 \pmod{p}$ this implies $ab + ba + \zeta b^2 = 0$. If $ab = b^2 = 0$, then (1) holds. If $b^2 = 0$ but $ab = \alpha p^{n-1}a \ne 0$, then replace a by $a' = \alpha a$, and (2) holds. Hence suppose $b^2 \ne 0$. Then $ab + ba + \zeta b^2 = 0$, all $\zeta \not\equiv 0 \pmod{p}$, implies $p = 2$,

$$b^2 = ab + ba = 2^{n-1}a.$$

If $ab = 0$ then (3) holds, so suppose $ab \ne 0$, $ba = 0$. Then, for $a' = a + b$, $a'^2 = a'b = 0$, $ba' = 2^{n-1}a'$, so again (3) holds. Clearly (1), (2), and (3) are not isomorphic.

Case II Choose $a \in S_1$, with $a^2 \ne 0$, and $b \in S_2$ with $ab = 0$. Then $a^2 = \beta p^i a$ for suitable integers $\beta \not\equiv 0 \pmod{p}$ and i, $1 \le i < n$. Replace a by $a' = \beta^{-1}a$, $\beta\beta^{-1} \equiv 1 \pmod{p^n}$, to obtain $a^2 = p^i a$. Let $ba = \alpha p^{n-1}a$. If $\alpha \not\equiv 0 \pmod{p}$ replace b by $\alpha^{-1}b$, where $\alpha\alpha^{-1} \equiv 1 \pmod{p}$. Hence we may assume $\alpha = 0$ or $\alpha = 1$. These lead directly to cases (1) and (2), and since (1) is commutative while (2) is not,

(1) and (2) describe non-isomorphic rings. To show that rings R in the three subcases of (1) are not isomorphic consider a general change of basis $a' = \zeta a + \eta b$, $b' = \kappa p^{n-1}a + \lambda b$, $\zeta \not\equiv 0$, $\lambda \not\equiv 0$ (mod p). It was shown above that $\alpha = 0$ and $\alpha \neq 0$ give non-isomorphic rings. If $\alpha \neq 0$ then $a'b' = 0$ implies $\eta \equiv 0$ (mod p), and then $a'^2 = p^i a'$ implies $\zeta \equiv 1$ (mod p). Then $b'^2 = \lambda^2 \alpha p^{n-1}a'$, so α can be changed only by a square, so $\alpha = 1$ and $a = \mu$ are disjoint. Finally, to show that the structure constant α in (2) cannot be changed in the case when $\alpha \neq 0$, consider a general change of basis of the above form. The condition $a'b' = 0$ then requires $\eta \equiv 0$ (mod p), and then $b'a' = p^{n-1}a'$ requires $\lambda \equiv 1$ (mod p). Then $a'^2 = p^i a'$ requires $\zeta \equiv 1$ (mod p), and then $b'^2 = \alpha p^{n-1}a'$, so α cannot be changed.

Case III In a similar way to Case II, we can choose $a \in S_1$ and $b \in S_2$ so that $a^2 = p^i a$, $1 \leq i < n$, and $ab = p^{n-1}a$. Let $b^2 = \alpha p^{n-1}a$, $0 \leq \alpha < p$.

First suppose $\alpha = 0$. Let $ba = \beta p^{n-1}a$, $0 \leq \beta < p$. Consideration of a general change of basis as above shows that β cannot be changed. Thus (1) holds.

Suppose $\alpha \neq 0$. Then $a - \alpha^{-1}b \in S_1$, where $\alpha\alpha^{-1} \equiv 1$ (mod p), and $(a - \alpha^{-1}b)b = 0$, so the hypothesis of Case III implies $(a - \alpha^{-1}b)^2 = 0$. Let $ba = \beta p^{n-1}a$, $0 \leq \beta < p$. Then $(a - \alpha^{-1}b)^2 = 0$ means

$$p^i a - \alpha^{-1}\beta p^{n-1}a = 0,$$

so $i = n - 1$ and $\beta = \alpha$. Consider a general change of basis $a' = \zeta(a + \eta b)$, $b' = \kappa p^{n-1}a + \lambda b$. Then

$$\begin{aligned} a'^2 &= \zeta^2(a^2 + \eta(ab + ba) + \eta^2 b^2) \\ &= \zeta^2(1 + \eta + \eta\beta + \eta^2\beta)p^{n-1}a \\ &= \zeta(1 + \eta\beta)(1 + \eta)p^{n-1}a'. \end{aligned}$$

The condition $a'^2 = p^{n-1}a'$ thus implies $1 + \eta\beta \not\equiv 0$ (mod p) and $1 + \eta \not\equiv 0$ (mod p). Further, the condition $a'b' = p^{n-1}a'$ implies $\lambda(1 + \eta\beta) \equiv 1$ (mod p). It follows that $b'a' = \beta'p^{n-1}a'$, where $\beta' \equiv \beta(1 + \eta)/(1 + \eta\beta)$ (mod p). By assumption $\beta = \alpha \neq 0$, which implies $\beta' \neq 0$, so cases (1) and (2) are disjoint. Moreover, if $\beta = 1$ then $\beta' = 1$, while for $\beta \neq 0$, $\beta \neq 1$, the expression β' takes on $p - 2$ values, i.e., all values except 0 and 1. Thus we can find η to force $\beta' = -1$, and the subcases of (2) are non-isomorphic.

Historical Remark The finite p-rings of rank 2 have been studied extensively by generators and relations in a series of papers by R. Ballieu [1, 2, 3] and by Ballieu and M. J. Schuind [1, 2]. Their results include in particular the classification of all rings of type $_p(n, 1)$.

2 The nilpotent algebras of dimension 4

The nilpotent, associative algebras of small dimensions have been studied for a long time. A list of those of dimension 4 was given by Allen [1] with corrections by Ghent [1], and a list of those of dimension 5 by Boyce [1]. None of these authors, however, settles the isomorphism problem. It would appear from Ghent [1], for example, that over a field of $q = p^n > 3$ elements, p an odd prime, there are

$$q^3 + 3q^2 + 3q + 2$$

not (directly) reducible, nilpotent, associative algebras of dimension 4, whereas the correct number is only $5q + 11$. In a very lengthy paper Scorza [7] again studies the algebras of dimension 4. He carefully formulates the isomorphism problem, but does not succeed in finding unique representatives of the isomorphism classes.

The purpose of the next several sections is to derive, in a very elementary, yet concise way, sets of generators and relations which uniquely define the non-isomorphic, nilpotent, associative algebras of dimension 4. Most of the complications arise in classifying the algebras with cube zero. Since associativity trivially holds for these algebras, one can easily describe sets of generators and relations which will define all the algebras in question. Settling the isomorphism problem, however, requires very careful selection of the generators and relations.

We shall denote the multiplicative group of the field $\mathfrak{F}$ by $\mathfrak{F}^*$, and denote by $\mathfrak{G}$ any fixed set of coset representatives of $(\mathfrak{F}^*)^2$ in $\mathfrak{F}^*$. Let us choose $\mathfrak{G}$ so that $1 \in \mathfrak{G}$. Note that if $\mathfrak{F}$ is finite, then $\mathfrak{G}$ contains exactly two elements if char $\mathfrak{F} \neq 2$, and one if char $\mathfrak{F} = 2$.

Suppose char $\mathfrak{F} = 2$. We shall denote by $\mathfrak{H}_1$ a fixed set of coset representatives of the additive subgroup $\{\varphi^2 + \varphi \mid \varphi \in \mathfrak{F}\}$ in the additive group of $\mathfrak{F}$. Further, the set $\mathfrak{S} = \{\varphi^2 \mid \varphi \in \mathfrak{F}\}$ is a subfield of $\mathfrak{F}$, and $\mathfrak{F}$ is divided into orbits under the linear fractional maps over $\mathfrak{S}$,

$$\varphi \to \frac{\alpha^2 + \beta^2 \varphi}{\gamma^2 + \delta^2 \varphi},$$

$\gamma^2 + \delta^2 \varphi \neq 0$, $\alpha^2 \delta^2 + \beta^2 \gamma^2 \neq 0$. We shall denote by $\mathfrak{H}_2$ a fixed set of representatives of these orbits. Note that when $\mathfrak{F}$ is finite, then $\mathfrak{H}_1$ contains two elements, and $\mathfrak{H}_2$ contains one element.

Henceforth Greek letters will denote field elements, and small Latin letters elements of the algebra A.

The results of these computations follow. All of the algebras described below, for each choice of the parameters indicated, are mutually non-isomorphic.

6.2.1 *If* $\dim A = 4$ *and* $A^3 \neq 0$, *then* A *has a basis* $\{a, a^2, a^3, b\}$ *which satisfies one of*:
(1) $b = a^4$, $a^5 = 0$ *(Power algebra)*.
(2) $\mathfrak{A}(A) \supseteq \langle a^3 \rangle$, $ab = 0$, $ba = \delta a^3$, $b^2 = \varepsilon a^3$, $\delta\,\varepsilon, = 0, 1$.

Remark   Case (2) with $\delta = \varepsilon = 0$ is reducible.

6.2.2 *If* $\dim A = 4$, $\dim A^2 = 1$, $A^3 = 0$, *and* A *is not reducible, then* A *has a basis* $\{a, b, c, d\}$ *with* $A^2 = \mathfrak{A}(A) = \langle d \rangle$ *such that* a, b, c *have one of the following multiplication tables.*

(1) $\begin{pmatrix} d & d & 0 \\ 0 & 0 & 0 \\ 0 & d & -d \end{pmatrix}$

(2) $\begin{pmatrix} d & d & 0 \\ 0 & \alpha d & d \\ 0 & d & 0 \end{pmatrix}$ *and* $\begin{array}{l} \textit{if char } \mathfrak{F} \neq 2 \textit{ then } \alpha = 0. \\ \textit{if char } \mathfrak{F} = 2 \textit{ then } \alpha \in \mathfrak{H}_1. \end{array}$

(3) $\begin{pmatrix} d & 0 & 0 \\ 0 & 0 & d \\ 0 & -d & 0 \end{pmatrix}$, *char* $\mathfrak{F} \neq 2$.

(4) $\begin{pmatrix} d & 0 & 0 \\ 0 & \alpha d & 0 \\ 0 & 0 & \beta d \end{pmatrix}$, $\begin{array}{l} \alpha, \beta \in \mathfrak{G} \textit{ are chosen so the quadratic forms} \\ x^2 + \alpha y^2 + \beta z^2 \textit{ form a set of representatives of the} \\ \textit{inequivalent ternary quadratic forms over } \mathfrak{F} \textit{ which} \\ \textit{have no non-trivial zeros.} \end{array}$

(5) $\begin{pmatrix} \gamma d & 0 & 0 \\ 0 & d & d \\ 0 & 0 & \varphi d \end{pmatrix}$, $\gamma \in \mathfrak{G}$, $\varphi \in \mathfrak{F}$.

Remark   Case (4) does not occur when $\mathfrak{F}$ is a finite field, since every ternary quadratic form over a finite field has a non-trivial zero (see, e.g., O'Meara [1], p. 158).

6.2.3 *If* $\dim A = 4$, $\dim A^2 = 2$, $A^3 = 0$, *and* A *is commutative, then* A *has a basis* $\{a, b, c, d\}$ *with* $a^2 = c$, $ab = ba = d$, $b^2 = \alpha c + \beta d$, *and*
If char $\mathfrak{F} \neq 2$, *then* $\beta = 0$, *and* $\alpha = 0$ *or* $\alpha \in \mathfrak{G}$.
If char $\mathfrak{F} = 2$, *then either*
(1) $\beta = 0$, $\alpha \in \mathfrak{H}_2$, *o* (2) $\beta = 1$, $\alpha \in \mathfrak{H}_1$.

Remark   For char $\mathfrak{F} \neq 2$, the algebra with $\alpha = 1$ is reducible. For char $\mathfrak{F} = 2$, the algebra with $\beta = 1$, $\alpha = 0$ is reducible.

6.2.4 *If* $\dim A = 4$, $\dim A^2 = 2$, $A^3 = 0$, *and* A *is not commutative, then* A *has a basis* $\{a, b, c, d\}$ *which satisfies one of the following conditions.*
(1) $a^2 = c$, $ab = 0$, $ba = d$, $b^2 = 0$.

(2) $a^2 = c$, $ab = d$, $b^2 = \gamma c$. If char $\mathfrak{F} \neq 2$, then $ba = -d$, and $\gamma = 0$ or $\gamma \in \mathfrak{G}$. If char $\mathfrak{F} = 2$, then $ba = c + d$ and $\gamma \in \mathfrak{H}_1$.

(3) $a^2 = c$, $ab = d$, $ba = \varphi d$, $b^2 = 0$, where $\varphi \in \mathfrak{F}$, $\varphi \neq \pm 1$.

In each of the following cases $a^2 = c$, $ab = d$, $ba = \alpha c + \beta d$, and $b^2 = \gamma c$, for suitable α, β, $\gamma \in \mathfrak{F}$.

(4) $\alpha = 1$, $\beta = \gamma = 0$.

(5) $\mathfrak{F}$ is the field of 3 elements. $\alpha = \beta = \gamma = 1$.

(6) Char $\mathfrak{F} \neq 2$. $\alpha = 1$, $\beta = 0$, $\gamma = 1$.

(7) Char $\mathfrak{F} \neq 2$. $\gamma \in \mathfrak{G}$, $\beta \in \mathfrak{F}$, $\beta \neq \pm 1$, and $\sigma = \alpha/(1 - \beta) \in \mathfrak{R}$, where $\mathfrak{R}$ is a minimal set of elements $\tau \in \mathfrak{F}$ such that the elements $\tau^2 - \gamma$ form a set of representatives of all cosets of $\mathfrak{F}^{*2}$ in $\mathfrak{F}$ which contain elements of the form $\tau^2 - \gamma$. In the case when $\beta \neq 0$ the algebra defined by α, β, γ and the algebra defined by $\alpha' = -\alpha/\beta$, $\beta' = 1/\beta$, $\gamma' = \gamma$ are isomorphic.

(8) Char $\mathfrak{F} = 2$. $\gamma \in \mathfrak{G}$, $\beta \neq 1$, $\sigma = \alpha/(1 + \beta) \in \mathfrak{F}$, and $1/(1 + \beta) \in \mathfrak{R}$, where $\mathfrak{R}$ is chosen as follows. For fixed σ, $\gamma \in \mathfrak{F}$ the elements of the form $(\zeta\sigma + \gamma)/(\zeta^2 + \gamma)$, $\zeta^2 \neq \gamma$, together with 0, form a subgroup $\mathfrak{H}_{\sigma,\gamma}$ of the additive group of $\mathfrak{F}$. $\mathfrak{R}$ is a set of coset representatives of $\mathfrak{H}_{\sigma,\gamma}$, where the representative of $\mathfrak{H}_{\sigma,\gamma}$ itself is chosen non-zero.

We now describe in turn the details of the classifications presented in each of **6.2.1** through **6.2.4**.

3 Assume dim $A = 4$, $A^3 \neq 0$

If dim $A^2 = 3$, then **4.1.4** implies that A is generated by a single element, and hence (1) of **6.2.1** holds. If dim $A^2 \leq 1$, then $A^3 = 0$, so we shall assume dim $A^2 = 2$, dim $A^3 = 1$, $A^4 = 0$.

Suppose $x^3 = 0$ for all $x \in A$. Then, since $A^3 \neq 0$, there exist elements a, b, linearly independent mod A^2, with one of a^2b, aba, or ab^2 not zero. First suppose $a^2b = d \neq 0$. Then $\{d, a^2\}$ is a basis for A^2. Let $ab = \alpha a^2 + \beta d$. Since $A^4 = 0$, $ad = 0$. Then $a^2b = a(ab) \neq 0$ implies $\alpha a^3 \neq 0$. Thus $a^3 \neq 0$. Second, suppose $aba = d \neq 0$. Then $\{ab, d\}$ is a basis for A^2. Let $ba = \alpha ab + \beta d$. Then
$$d = aba = a(ba) = \alpha a^2b \neq 0,$$
so the first case holds. The third case, $ab^2 \neq 0$, is dual to the first. Hence A always contains an element a with $a^3 \neq 0$.

Let $\{a, a^2, a^3, b\}$ form a basis for A. For $ab = \alpha a^2 + \beta a^3$, replacing b by $b' = b - \alpha a - \beta a^2$ forces $ab = 0$. Let $b^2 = \alpha a^2 + \beta a^3$. Then $0 = (ab)b = ab^2 = \alpha a^3$ so $\alpha = 0$. Finally let $ba = \gamma a^2 + \delta a^3$. Then $0 = (ab)a = a(ba) = \gamma a^3$, so $\gamma = 0$. Thus we assume $ab = 0$, $ba = \delta a^3$, $b^2 = \varepsilon a^3$.

To complete the proof that (2) of **6.2.1** holds we must show that δ and ε may be forced 0 or 1. If $\delta \neq 0$, then replacing b by $b' = \delta^{-1}b$ forces $\delta = 1$. For $\delta = 0$ and $\varepsilon \neq 0$, replacing a by $a' = \varepsilon a$ and b by $b' = \varepsilon b$ forces $\varepsilon = 1$. For $\delta = 1$ and $\varepsilon \neq 0$, finally, replacing a by $a' = \varepsilon^{-1}a$ and b by $b' = \varepsilon^{-2}b$ forces $\varepsilon = 1$. Thus (2) of **6.2.1** holds.

The four algebras described in (2) of **6.2.1** are mutually non-isomorphic, since $\delta = 0$ or $\delta = 1$ according as A is commutative or not, and $\varepsilon = 1$ or $\varepsilon = 0$ according as there exists $x \in A$ with $x^2 \neq 0$, $x^3 = 0$, or not.

4 Assume dim $A = 4$, dim $A^2 = 1$, $A^3 = 0$, and A is not reducible

6.4.1 *A contains an element x with $x^2 \neq 0$.*

Proof (Boyce [1])   If $x^2 = 0$ for all $x \in A$, then it is easy to construct a basis $\{a, b, c, d\}$ for A with $ab = -ba = ac = -ca = d$,
$$bc = -cb = \alpha d,$$
some $\alpha \in \mathfrak{F}$. Let $c' = c + \alpha a - b$. Then
$$ac' = c'a = bc' = c'b = c'^2 = 0,$$
so A is reducible.

Let us for the present consider an $(n + 1)$-dimensional algebra N, $n \geq 3$, over $\mathfrak{F}$ with dim $N^2 = 1$, $N^3 = 0$, and N not reducible. Let $N^2 = \langle f \rangle$ and extend $\{f\}$ to a basis $\{e_1, \ldots, e_n, f\}$ for N. Let $e_i e_j = m_{ij}f$, $1 \leq i, j \leq n$, $m_{ij} \in \mathfrak{F}$. By **6.4.1** we can assume $m_{11} = 1$. By replacing, then, e_j by $e_j' = e_j - m_{j1}e_1$, we force $m_{j1} = 0$, $2 \leq j \leq n$. If any $m_{1j} \neq 0$, $2 \leq j \leq n$, then by replacing e_j by e_2 and e_2 by $m_{1j}^{-1}e_j$ we obtain $m_{12} = 1$. By then replacing e_j by $e_j' = e_j - m_{1j}e_2$ we obtain $m_{1j} = 0$, $3 \leq j \leq n$. Thus in any case the algebra N is represented by an $n \times n$ matrix $M = (m_{ij})$ which has the block form

$$M = \left(\begin{array}{c|c} 1 & \varepsilon \ 0 \ldots 0 \\ \hline 0 & \\ \vdots & B \\ 0 & \end{array} \right)$$

where $\varepsilon = 0$ or $\varepsilon = 1$.

We now suppose $\varepsilon = 1$ and seek to find a change of basis of N which makes $\varepsilon' = 0$ while maintaining the block form indicated. A change of basis of N defines a nonsingular $n \times n$ matrix T over $\mathfrak{F}$ and takes the matrix M of structure constants onto $M' = T^t M T$, where T^t denotes the transpose of T, and the image of e_i is in column i of

T, $1 \leq i \leq n$. Let us write T in the block form

$$T = \left(\frac{w \mid X}{Y \mid Z}\right).$$

The conditions that M' have the desired form, with $\varepsilon' = 0$, are then

$$wX^t + y_1 X^t + Z^t BY = (0 \ldots 0)^t \qquad (1)$$

and
$$Xw + wZ_1 + Y^t BZ = (0 \ldots 0), \qquad (2)$$

where y_1 denotes the first entry of the column vector Y, and Z_1 the first row of Z. Subtracting (1)^t from (2) we obtain

$$[-y_1]X + [w]Z_1 + [Y^t(B - B^t)]Z = (0 \ldots 0).$$

Surely $\qquad [-y_1]w \mid [w]y_1 + [Y^t(B - B^t)]Y = 0.$

Hence we have a linear dependence among the rows of T. Thus

$$y_1 = 0 \text{ and } Y^t(B - B^t) + (w0 \ldots 0) = (0 \ldots 0). \qquad (3)$$

We now return to the special case $n = 3$, and let $N = A$. We shall show that $\varepsilon' = 0$ is impossible if and only if A has a basis satisfying (1) or (2) of **6.2.2**.

Condition (3) for forcing $\varepsilon' = 0$ becomes

$$y_1 = 0 \text{ and } w + y_2(m_{32} - m_{23}) = 0, \qquad (3a)$$

and (1)^t becomes

$$wX + Y^t B^t Z = (0\ 0). \qquad (1a)$$

Hence it is possible to force $\varepsilon' = 0$ if and only if there is a non-singular matrix T which satisfies (1a) and (3a). We shall distinguish two cases according to $m_{32} - m_{23}$ zero or not.

Case A: $m_{32} \neq m_{23}$ Replacing e_3 by $e_3' = (m_{23} - m_{32})^{-1}e_3$ we obtain $m_{23} - m_{32} = 1$, so by (3a) $w = y_2$. Since we only seek to make $\varepsilon' = 0$, without loss of generality we may assume that $w = y_2 = 1$. Condition (1a) then becomes $X = -(m_{23} m_{33})Z$, and implies det T $= (\det Z) (1 + m_{33})$. Hence we can find T which forces $\varepsilon' = 0$ if and only if $m_{33} \neq -1$.

To characterize the algebras with $m_{33} = -1$ and $m_{23} = m_{32} + 1$, we first replace e_1 by $e_1' = -e_1$ and e_2 by $e_2' = -e_2 - m_{23}e_3$. This makes $m_{23} = 0$, $m_{32} = 1$. For $m_{22} = 0$ we obtain case (1) of **6.2.2**. For $m_{22} \neq 0$ define $e_1' = e_2$, $e_2' = e_2 - m_{22}e_1$, $e_3' = e_1 - e_3$, and $f' = m_{22}f$. Then $m_{32}' = m_{23}'$, and the algebra enters Case B.

Case B: $m_{32} = m_{23}$ Condition (3a) for forcing $\varepsilon' = 0$ becomes $w = y_1 = 0$. Nonsingularity of T then requires $y_2 \neq 0$. Condition (1a) becomes $(m_{23} m_{33}) Z = (0\ 0)$. Hence for $m_{23} = 0$ we can force $\varepsilon' = 0$ with

$$T = \begin{pmatrix} 0 & 1 & 0 \\ 0 & 0 & 1 \\ 1 & 0 & 0 \end{pmatrix},$$

and, for $m_{33} \neq 0$, we can force $\varepsilon' = 0$ with

$$T = \begin{pmatrix} 0 & 1 & 0 \\ 0 & m_{33} & m_{33} \\ 1 & -m_{23} & -m_{23} \end{pmatrix}.$$

Thus suppose $m_{23} \neq 0$ and $m_{33} = 0$. Condition (1a) then requires that the second row of T be 0, so we cannot force $\varepsilon' = 0$. Replacing e_3 by $e_3' = m_{23}^{-1} e_3$ we obtain $m_{23} = 1$. If char $\mathfrak{F} \neq 2$, replace e_2 by $e_2' = e_2 - \frac{1}{2} m_{22} e_3$. This forces $m_{22} = 0$, so (2) of **6.2.2** holds. If char $\mathfrak{F} = 2$, replace e_1 by $e_1' = e_1 + \varphi e_3$ and e_2 by $e_2' = e_2 + \varphi e_1$, $\varphi \in \mathfrak{F}$. Then m_{22} is replaced by $m_{22}' = \varphi^2 + \varphi + m_{22}$. Hence we can force $m_{22} \in \mathfrak{H}_1$. Thus (2) of **6.2.2** holds.

Since case (1) of **6.2.2** has a singular matrix M, while case (2) has a nonsingular M, cases (1) and (2) describe non-isomorphic algebras. The center of an algebra satisfying (2) with char $\mathfrak{F} = 2$ is $\langle e_3, f \rangle$. A direct check then shows that the subcases of (2) are non-isomorphic.

We conclude this section by determining the algebras A for which we can force $\varepsilon' = 0$. For such an algebra an appropriate change of basis gives

$$M = \left(\begin{array}{c|cc} \gamma & 0 & 0 \\ \hline 0 & & \\ 0 & & B \end{array} \right)$$

where $\gamma \in \mathfrak{G}$ and B is in one of the standard forms for an algebra of dimension 3 given in (2), (4), or (5) of **2.3.6**. These lead in an obvious way to cases (3), (4), and (5) of **6.2.2**. An algebra A which would be described by (3) of **6.2.2** with char $\mathfrak{F} = 2$ is, however, isomorphic to an algebra of case (4) under the change of basis $a' = a + b$, $b' = a + c$, $c' = a + b + c$.

We have already shown that algebras of cases (1) and (2) of **6.2.2** are not isomorphic to algebras of cases (3), (4), (5). Commutativity conditions imply immediately that (3), (4), and (5) are mutually non-isomorphic.

5 Assume dim $A = 4$, dim $A^2 = 2$, $A^3 = 0$

First suppose A is commutative. Since dim $A^2 = 2$, it is easy to find elements a, $b \in A$, linearly independent mod A^2, for which $a^2 = c$ and $ab = ba = d$ are linearly independent. Thus $A^2 = \langle c, d \rangle$, and there are elements α, $\beta \in \mathfrak{F}$ such that $b^2 = \alpha c + \beta d$.

Suppose char $\mathfrak{F} \neq 2$. Replace b by $b' = b - (\beta/2)a$ and d

by $d' = d - (\beta/2)c$. Then β is replaced by 0. Replacing b by an appropriate multiple of itself then forces $\alpha = 0$ or $\alpha \in \mathfrak{G}$. Hence **6.2.3** holds.

Suppose char $\mathfrak{F} = 2$ and $\beta = 0$. For $\kappa, \lambda \in \mathfrak{F}$, $\lambda \neq 0$, the change of basis $b' = \kappa a + \lambda b$, $d' = \kappa c + \lambda d$, $\lambda \neq 0$, makes $\alpha' = \kappa^2 + \lambda^2 \alpha$, so an appropriate choice of κ, λ forces $\alpha' \in \mathfrak{H}_2$. Thus (1) of **6.2.3** holds.

Suppose char $\mathfrak{F} = 2$ and $\beta \neq 0$. First replace b by $b' = \beta^{-1}b$ and d by $d' = \beta^{-1}d$. Then β is replaced by 1. Next, for $\varphi \in \mathfrak{F}$, define $b' = b + \varphi a$ and $d' = d + \varphi c$. Then $\beta' = 1$ and $\alpha' = \alpha + \varphi^2 + \varphi$, so an appropriate choice of φ forces $\alpha' \in \mathfrak{H}_1$. Thus (2) of **6.2.3** holds. A direct check shows that (1) and (2) and their subcases are not isomorphic.

For the remainder of this section we assume A is not commutative.

6.5.1 *If A is not commutative, then it has a basis $\{a, b, c, d\}$ which satisfies either* (1) *of* **6.2.4** *or*
$$a^2 = c, \ ab = d, \ ba = \alpha c + \beta d, \ b^2 = \gamma c + \delta d \qquad \text{(i)}$$
for suitable $\alpha, \beta, \gamma, \delta \in \mathfrak{F}$. These two cases are disjoint.

Proof Surely there exists $a \in A$ with $a^2 \neq 0$. Let $a^2 = c \neq 0$. Choose $b \in A$, $b \notin \langle a, A^2 \rangle$. Then $ab \in \langle c \rangle$, or else (i) holds for $d = ab$. Assume $ab \in \langle c \rangle$. For the first case suppose $b^2 = d \notin \langle c \rangle$. Then $ba \in \langle d \rangle$, or else (i) holds for $a' = b$, $b' = a$, $c' = d$, $d' = ba$. Assume $ba \in \langle d \rangle$. Let $ab = \varphi c$, $ba = \psi d$. If $\varphi\psi \neq 1$, then (i) holds for $a' = a + b$, $b' = b$, $c' = (1 + \varphi)c + (1 + \psi)d$, $d' = \varphi c + d$. If $\varphi\psi = 1$, then replacing b by $b' = a - \psi b$ moves A into the second case, which is $b^2 \in \langle c \rangle$. In the second case $ba = d \notin \langle c \rangle$ follows from dim $A^2 = 2$. Further, $b^2 = 0$, or else (i) holds for $a' = b$, $b' = a$, $c' = b^2$, $d' = d$. Moreover, $ab = 0$, or else (i) holds for $a' = a + b$, $b' = b$, $c' = c + ab + d$, $d' = ab$. Hence (1) of **6.2.4** holds. A direct check shows that the algebra (1) of **6.2.4** has no basis satisfying (i).

6.5.2 *If A has a basis which satisfies* (i) *and A is not commutative, then the basis may be chosen so that*
$$a^2 = c, \ ab = d, \ ba = \alpha c + \beta d, \ b^2 = \gamma c. \qquad \text{(ii)}$$

Proof Assume A has a basis $\{a, b, c, d\}$ which satisfies (i) with $\delta \neq 0$. Then (ii) holds for a basis $\{a', b', c', d'\}$ of A chosen as follows:
If $\beta \neq -1$ let $a' = a$, $b' = b - \delta(\beta + 1)^{-1}a$, $c' = c$,
$$d' = d - \delta(\beta + 1)^{-1}c.$$
If $\beta = -1$ and $\alpha \neq 0$, let $a' = \alpha a - b$, $b' = b$, $c' = \gamma c + \delta d$, $d' = -\gamma c + (\alpha - \delta)d$.
If, finally, $\beta = -1$ and $\alpha = 0$, then char $\mathfrak{F} \neq 2$ since A is not

H

commutative. Let $a' = b - \delta a$, $b' = b + \delta a$, $c' = (\gamma + \delta^2)c + \delta d$, $d' = (\gamma - \delta^2)c - \delta d$.

The cases $\beta = -1$ and $\beta \neq -1$ in bases satisfying (ii) lead to non-isomorphic algebras, since for $\beta = -1$ all squares are multiples of the fixed element c, while for $\beta \neq -1$ there are elements (e.g., a and $a + b$) with linearly independent squares.

6.5.3 *Suppose A is not commutative, and has a basis $\{a, b, c, d\}$ which satisfies* (ii) *with $\beta = -1$. Then A has a basis which satisfies* (2) *of* **6.2.4**. *The subcases of* (2) *are mutually non-isomorphic.*

Proof First suppose char $\mathfrak{F} \neq 2$. Replace b by $b' = b - (\alpha/2)a$ and d by $d' = d - (\alpha/2)c$. Then the basis $\{a, b, c, d\}$ satisfies (ii) with $\alpha = 0$, $\beta = -1$. For $\varphi \in \mathfrak{F}$, $\varphi \neq 0$, replacing b by φb and d by φd forces $\gamma = 0$ or $\gamma \in \mathfrak{G}$. Hence (2) of **6.2.4** holds. A direct check shows that different values of γ give non-isomorphic algebras.

Suppose char $\mathfrak{F} = 2$. Since A is not commutative, the basis satisfying (ii) has $\alpha \neq 0$. Replace b by $b' = \alpha^{-1}b$ and d by $d' = \alpha^{-1}d$, and (ii) holds with $\alpha = \beta = 1$. Let $b' = b + \varphi a$, $d' = d + \varphi c$, $\varphi \in \mathfrak{F}$. Then $\{a, b', c, d'\}$ satisfies (ii) with $\alpha' = \beta' = 1$, $\gamma' = \varphi^2 + \varphi + \gamma$. Hence we can force $\gamma' \in \mathfrak{H}_1$. Thus (2) of **6.2.4** holds. Again, a direct computation shows that different values of γ define non-isomorphic algebras.

Henceforth we suppose that A has a basis $\{a, b, c, d\}$ which satisfies (ii) with $\beta \neq -1$. Let us study a general change of basis of A given by
$$a' = \zeta a + \eta b, \ b' = \kappa a + \lambda b, c' = a'^2, d' = a'b'.$$
Nonsingularity requires
$$\zeta\lambda - \eta\kappa \neq 0. \tag{iii}$$
We wish the basis $\{a', b', c', d'\}$ to satisfy (ii); hence we wish $b'^2 = \gamma'a'^2$ for some $\gamma' \in \mathfrak{F}$. This requires
$$\kappa\lambda = \gamma'\zeta\eta \text{ and } \kappa^2 + \lambda^2\gamma = \gamma'(\zeta^2 + \eta^2\gamma). \tag{iv}$$
(iv) gives $\zeta\eta(\kappa^2 + \lambda^2\gamma) = \gamma'\zeta\eta(\zeta^2 + \eta^2\gamma) = \kappa\lambda(\zeta^2 + \eta^2\gamma)$, so
$$(\eta\lambda\gamma - \kappa\zeta)(\zeta\lambda - \eta\kappa) = 0,$$
so (iii) gives $\eta\lambda\gamma - \kappa\zeta = 0$. This, with (iv), gives $\gamma' = 0$ if and only if $\gamma = 0$. Hence the cases $\gamma = 0$ and $\gamma \neq 0$ give non-isomorphic algebras.

Suppose $\gamma = 0$. A direct check shows that the cases $\alpha = 0$ and $\alpha \neq 0$ lead to non-isomorphic algebras, and that, for $\alpha = 0$, the structure constant β cannot be changed within an isomorphism class, so that (3) of **6.2.4** holds. If $\alpha \neq 0$, let $\zeta = \alpha$, $\eta = \beta$, $\kappa = 0$, $\lambda = 1 + \beta$. This gives a basis satisfying (ii) with $\alpha' = 1$, $\beta' = 0$, $\gamma' = 0$. Thus (4) of **6.2.4** holds.

Henceforth we assume $\gamma \neq 0$. The condition $\eta\lambda\gamma - \kappa\zeta = 0$ together with (iv) implies that $\gamma\gamma' = \varphi^2$ for some $\varphi \in \mathfrak{F}$, and also that $\kappa = \varphi\eta$ and $\lambda = \varphi\zeta/\gamma$. For $\zeta = 1$, $\eta = \kappa = 0$, we obtain $\gamma' = \lambda^2\gamma$. Hence we can force $\gamma' \in \mathfrak{G}$, and, since $\gamma\gamma'$ is always a square, different members of $\mathfrak{G}$ lead to non-isomorphic algebras.

Henceforth we assume $\gamma = \gamma' \in \mathfrak{G}$. Then, for $\varepsilon = \pm 1$, we have $\kappa = \varepsilon\gamma\eta$ and $\lambda = \varepsilon\zeta$. Conditions for the nonsingularity of the transformation on $\langle a, b \rangle$ and the induced transformation on $\langle c, d \rangle$ become

$$\zeta^2 - \eta^2\gamma \neq 0 \text{ and } \delta = \zeta^2 + \zeta\eta\alpha - \eta^2\beta\gamma \neq 0. \tag{v}$$

An elementary but somewhat tedious computation gives

$$\alpha' = \varepsilon[2\zeta\eta\gamma(1 - \beta) + \alpha(\zeta^2 + \eta^2\gamma)]/\delta \tag{vi}$$

and

$$\beta' = [\zeta^2\beta - \zeta\eta\alpha - \eta^2\gamma]/\delta. \tag{vii}$$

6.5.4 *Under the above assumptions, either a basis for A may be chosen to satisfy (ii) with $\beta \neq +1$, or else A satisfies (5) of* **6.2.4**.

Proof Suppose $\beta = 1$ for every basis of A satisfying (ii). Since A is not commutative, $\alpha \neq 0$. Since $\beta \neq -1$, char $\mathfrak{F} \neq 2$. Then $\beta' = \beta = 1$ implies by (vii) that $\zeta\eta = 0$ for all ζ, η satisfying (v). If $\mathfrak{F}$ has more than 3 elements, however, then there exists $\zeta \in \mathfrak{F}$ such that $\zeta \neq 0$, $\zeta^2 \neq \gamma$, and $\zeta^2 + \zeta\alpha \neq \gamma$. This ζ, together with $\eta = 1$, satisfies (v) and makes $\beta' \neq 1$. Thus $\mathfrak{F}$ is the field of 3 elements. If $\gamma \neq 1$ then $|\mathfrak{F}| = 3$ implies $\gamma = -1$, so $\zeta = -\alpha$, $\eta = 1$ satisfies (v) and makes $\beta' \neq 1$. Thus $\gamma = 1$. By replacing b by $b' = \alpha b$ and d by $d' = \alpha d$, finally, α' becomes 1, and so A satisfies (5) of **6.2.4**.

From now on we shall assume $\beta \neq 1$, $\beta' \neq 1$. Let $\sigma = \alpha/(1 - \beta)$ and $\sigma' = \alpha'/(1 - \beta')$. A direct calculation gives

$$\sigma'^2 - \gamma = \left(\frac{\zeta^2 - \eta^2\gamma}{\zeta^2 + 2\zeta\eta\sigma + \eta^2\gamma}\right)^2 (\sigma^2 - \gamma). \tag{viii}$$

Note that by (viii) and (v) the cases $\sigma^2 = \gamma$ and $\sigma^2 \neq \gamma$ are disjoint, and, for char $\mathfrak{F} = 2$, that $\sigma' = \sigma$, while for char $\mathfrak{F} \neq 2$, the quantity $\sigma^2 - \gamma$ can be altered only by a square. Henceforth, we must consider the cases char $\mathfrak{F} \neq 2$ and char $\mathfrak{F} = 2$ separately.

6.5.5 *Suppose $\sigma^2 = \gamma$ and char $\mathfrak{F} \neq 2$. Then A satisfies (6) of* **6.2.4**.

Proof Since $\gamma = \sigma^2$ is a square in $\mathfrak{G}$, by our choice of $\mathfrak{G}$ we have $\gamma = \sigma^2 = 1$. Hence $\alpha = \sigma(1 - \beta)$ and $\sigma = 1$ or $\sigma = -1$. Let $\zeta = \sigma$, $\eta = \beta$, $\varepsilon = \sigma$. Then (v) holds, $\alpha' = 1$, and $\beta' = 0$, so A satisfies (6) of **6.2.4**.

6.5.6 *Suppose $\sigma^2 \neq \gamma$ and char $\mathfrak{F} \neq 2$. Then A satisfies (7) of* **6.2.4**.

Proof First we determine ζ, η, ε so that $\sigma' = \tau$, where $\tau^2 - \gamma$ is a predetermined representative of the coset of $(\mathfrak{F}^*)^2$ in $\mathfrak{F}^*$ which contains $\sigma^2 - \gamma$. Let $\tau^2 - \gamma = \varphi^2(\sigma^2 - \gamma)$, $\varphi \in \mathfrak{F}$, $\varphi \neq 0$. If $\sigma = \tau$, then we are through. Otherwise, by the quadratic formula and condition (viii), $\sigma' = \tau$ when $\zeta = \omega_2\tau - \omega_1\varphi\sigma$, $\eta = \omega_1\varphi - 1$, and $\varepsilon = -\omega_2$, with $\omega_1 = \pm 1$ and $\omega_2 = \pm 1$. For this choice of ζ and η, moreover, condition (v) becomes $\omega_1\varphi(\omega_2\sigma\tau - \gamma) \neq \tau^2 - \gamma$ and $\varphi - \omega_1 \neq 2\varphi/(1 - \beta)$. It is easy to check that at least one of the four choices for ω_1, ω_2 satisfies both inequalities.

Finally, if $\sigma' = \sigma$, then a direct calculation, given (vi) and (vii), implies that the only further changes of basis possible occur when $\beta \neq 0$, and satisfy $\alpha' = -\alpha/\beta$, $\beta' = 1/\beta$. Such a change occurs for $\zeta = 0$, $\eta = 1$, $\varepsilon = 1$. Hence A satisfies (7) of **6.2.4**, and the algebras described in (7) are mutually non-isomorphic.

6.5.7 *Suppose* char $\mathfrak{F} = 2$. *Then A satisfies* (8) *of* **6.2.4**.

Proof By (viii) $\sigma' = \sigma$, and so the structure constant σ is an invariant of the algebra A. Moreover, by (vii),

$$\frac{1}{1 + \beta'} = \frac{1}{1 + \beta} + \frac{\zeta\eta\sigma + \eta^2\gamma}{\zeta^2 + \eta^2\gamma}.$$

For fixed γ, σ, the set

$$\mathfrak{H}_{\sigma, \gamma} = \{(\zeta\eta\sigma + \eta^2\gamma)/(\zeta^2 + \eta^2\gamma) \mid \zeta, \eta \in \mathfrak{F}, \zeta^2 + \eta^2\gamma \neq 0\}$$

is an additive subgroup of $\mathfrak{F}$. Hence $1/(1 + \beta)$ may be chosen to be a predetermined coset representative of $\mathfrak{H}_{\sigma, \gamma}$, subject to the conditions in (v). The condition $\delta \neq 0$ is equivalent to choosing a non-zero representative of $\mathfrak{H}_{\sigma, \gamma}$ itself. The other condition, $\zeta^2 + \eta^2\gamma \neq 0$, holds by definition of $\mathfrak{H}_{\sigma, \gamma}$. Finally, if $\beta' = \beta$, then $\sigma' = \sigma$ implies $\alpha' = \alpha$, so no further change is possible. Thus A satisfies (8) of **6.2.4**, and the algebras described are mutually non-isomorphic.

Bibliography

In ADDITION to the works cited in the text, the bibliography contains all papers available to the authors bearing to a non-trivial extent on the structure of nilpotent rings. After some of the papers there is a reference to a section or result in the text, indicating that some (but not necessarily all) of the results in the paper are discussed there. After other papers are a few words of description, meant not to summarize the papers, but only, in conjunction with the title, to help the reader locate the general context in which the paper is placed. The absence of a comment after any paper carries no particular significance. We have included, whenever possible, references to one of the reviewing journals. *Mathematical Reviews* is denoted by MR, *Zentralblatt für Mathematik* by ZM, and *Dissertation Abstracts* by DA.

ADO, I. D.
1. On nilpotent algebras and p-groups, *C. R. (Doklady) Acad. Sci. URSS* (*N.S.*), **40** (1943), 299–301; MR **6**: 146. Connections between locally finite p-groups and locally nilpotent p-algebras via modular group algebras and the circle composition.

ALBERT, A. A.
1. *Structure of Algebras*, Amer. Math. Soc. Colloq. Publ., **24**, 1939; Revised, 1961; MR **1**: 99.

ALLEN, R. B.
1. On hypercomplex number systems belonging to an arbitrary domain of rationality, *Trans. Amer. Math. Soc.*, **9** (1908), 203–218. Criteria for reducibility of algebras. List of algebras of dimensions at most 4. Not entirely correct. See Ghent [1].

ANDRIJANOV, V. I.
1. Mixed Hamiltonian nilrings (Russian), *Ural. Gos. Univ. Mat. Zap.*, **5** (1966), tetrad' 3, 15–30; MR **34**: 1359. Rings with mixed additive group in which all subrings are ideals.
2. Periodic Hamiltonian rings (Russian), *Mat. Sb. (N.S.)*, **74 (116)** (1967), 241–261; MR **36**: 212. See §(4.6).

ANDRIJANOV, V. I., FREĬDMAN, P. A.
1. Hamiltonian rings (Russian), *Sverdlovsk. Gos. Ped. Inst. Učen. Zap.*, **31** (1965), 3–23; MR **35**: 5469. Study of several classes of rings in which all subrings are ideals.

BALLIEU, R.
1. Anneaux finis; systèmes hypercomplexes de rang deux sur un corps, *Ann. Soc. Sci. Bruxelles* (1), **61** (1947), 117–126; MR **8**: 499. Cyclic p-rings, algebras of dimension 2.
2. Anneaux finis; systèmes hypercomplexes de rang trois sur un corps commutatif, *Ann. Soc. Sci. Bruxelles* (1), **61** (1947), 222–227; MR **9**: 267. List of the rings of order p^3. No proofs.
3. Anneaux finis à module de type (p, p^2), *Ann. Soc. Sci. Bruxelles* (1), **63** (1949), 11–23; MR **11**: 711. Complete determination of rings of type $_p(2, 1)$.

BALLIEU, R., SCHUIND, M.-J.
1. Anneaux finis à module de type (p, p^r), *Ann. Soc. Sci. Bruxelles* (1), **63** (1949), 137–147; MR **11**: 711. See §(6.1).
2. Anneaux à module de type (p^m, p^{m+n}), *Ann. Soc. Sci. Bruxelles* (1), **65** (1951), 33–40; MR **12**: 796. Determination of some special classes of rings of type $_p(m, m + n)$.

BAUMGARTNER, K.
1. Bemerkungen zum Isomorphieproblem der Ringe, *Monatsh. Math.*, **70** (1966), 299–308; MR **35**: 216. Formulation of isomorphism in terms of linear transformations in Kronecker product.

BEAUMONT, R. A.
1. Rings with additive group which is the direct sum of cyclic groups, *Duke Math. J.*, **15** (1948), 367–369; MR **10**: 10. Defining rings by introducing a product on a basis. Cyclic rings.

BECHTELL, H.
1. Nongenerators of rings, *Proc. Amer. Math. Soc.*, **19** (1968), 241–245. Two-sided Frattini subring and its relations to the radical of a ring.

BLOOM, D. M.
1. List of the 11 rings of order 4, *Amer. Math. Monthly*, Elementary problem, **71** (1964), 918–920.

BOERS, A. H.
1. L'anneau à quatre éléments, *Nederl. Akad. Wetensch. Proc. Ser. A* **69** = *Indag. Math.* **28** (1966), 14–21; MR **33**: 156. List of rings, not necessarily associative, with four elements.

BOYCE, F. W.
1. Certain types of nilpotent algebras, Ph.D. dissertation, Univ. of Chicago, 1938. Constructs nilpotent algebras of dimension 5 over a field of characteristic not 2.

BRAMERET, M. P.
1. Treillis d'idéaux et structure d'anneaux, *C. R. Acad. Sci. Paris (A-B)*, **255** (1962), 1434–1435; MR **26**: 140. Characterizes nilpotent rings with totally ordered lattice of ideals.

BÜKE, A.
1. Untersuchungen über kommutativ-assosiativ und nilpotenten Algebren von Index 3 and von der Charakteristik 2, *Rev. Fac. Sci. Univ. Istanbul Sér. A*, **19** (1954), Suppl. 1–145; MR **16**: 561.

2. Nilpotente Algebren vom Index 3 über einen Körper K der Charakteristik 2, *Rev. Fac. Sci. Univ. Istanbul (A)*, **22** (1957), 45–89, 110; MR **20**: 5221. Study of the commutative algebras A described in the title, by means of studying the possible identifications which can occur in the lattice of ideals of A generated by A and $Q = \{x \in A \mid x^2 = 0\}$, and closed under the operations $+$, $\cap$, ring product, and taking annihilators.

BURNSIDE, W.
1. *Theory of Groups of Finite Order*, 2nd ed., Cambridge, 1911. Reprinted, Dover, 1955.

CADA, L. J.
1. A structure theory for linear associative nilpotent algebras, Ph.D. dissertation, Cath. Univ. of America, 1964; DA **26**: 1058. Study of certain annihilator series, extension theory for nilpotent algebras.

CARTAN, E,
1. Les groupes bilinéaires et les systèmes de nombres complexes, *Ann. Fac. Sci. Univ. Toulouse*, **12B** (1898), 1–99. Structure theory for finite-dimensional semi-simple algebras over the complex numbers.

CAYLEY, A.
1. On double algebra, *Proc. London Math. Soc.* (1), **15** (1883), 185–197. Not necessarily associative algebras of dimension 2.

CHARLES, B.
1. Sur l'algèbre des opérateurs linéaires, *J. Math. Pures Appl.* (9) **33** (1954), 81–145; MR **14**: 768, 721, 939; **16**: 439. Maximal commutative algebras of matrices.

COLEMAN, D. B.
1. On the modular group ring of a p-group, *Proc. Amer. Math. Soc.* **15** (1964), 511–514; MR **29**: 2306. Studies group of units and center of a modular group algebra of a p-group.

DESKINS, W. E.
1. Finite Abelian groups with isomorphic group algebras, *Duke Math. J.*, **23** (1956), 35–40; MR **17**: 1052. Modular group algebras of finite abelian p-groups are isomorphic if and only if the groups are isomorphic.

DEURING, M.
1. *Algebren*, Springer Verlag, 1935; ZM **11**: 198.

DICKSON, L. E.
1. *Linear Groups, with an Exposition of the Galois Field Theory*, Leipzig, 1901. Reprinted, Dover, 1958.
2. Modular theory of group-matrices, *Trans. Amer. Math. Soc.*, **8** (1907), 389–398. See **2.4.1**.
3. *Linear Algebras*, Cambridge, 1914. Reprinted, Hafner, 1960.
4. *Algebras and their Arithmetics*, Univ. of Chicago Press, 1923. Reprinted, Dover, 1960.

DIECKMANN, E. M.
1. Isomorphism of group algebras of p-groups, Ph.D. dissertation, Washington Univ., 1967; DA **28**: 4194B. In the modular case studies mainly p-groups of class 2.

DIVINSKY, N.
1. Pseudo-regularity, *Canadian J. Math.* **7** (1955), 401–410; MR **17**: 8. Generalization of circle composition.
2. *Rings and Radicals*, Univ. of Toronto Press, 1965; MR **33**: 5654.

DORROH, J. L.
1. Concerning adjunctions to algebras, *Bull. Amer. Math. Soc.*, **38** (1932), 85–88; ZM **3**: 387. Adjoining identities via ordered pairs.

DUBISCH, R., PERLIS, S.
1. On total nilpotent algebras, *Amer. J. Math.*, **73** (1951), 439–452; MR **12**: 798. See §§(2.1), (4.2).

DYMENT, Z. M.
1. Maximal commutative nilpotent subalgebras of a matrix algebra of the sixth degree (Russian), *Vesci Akad. Navuk BSSR Ser. Fiz.-Mat. Navuk*, **1966**, No. 3, 53–68; MR **34**: 2622. Finds all such mutually non-conjugate subalgebras of a 6×6 complete matrix algebra over an algebraically closed field of characteristic 0. The number of these is finite.

EGGERT, N. H., Jr.
1. Some results on finite rings whose groups of units are abelian, Ph.D. dissertation, Univ. of Colorado, 1966; DA **28**: 2515B. Structure of group of units of a finite commutative algebra.
2. Quasi-regular groups of finite commutative nilpotent algebras. To appear.

ELDRIDGE, K. E.
1. Descending chain condition rings with cyclic quasi-regular group, Ph.D. dissertation, Univ. of Colorado, 1965; DA **26**: 6736.
2. On ring structures determined by groups. To appear. Implications for an Artinian ring of conditions on its group of quasi-regular elements.
3. Ring structure determined by groups. II. To appear. Artinian rings with nilpotent group of quasi-regular elements.

ELDRIDGE, K. E., FISCHER, I.
1. D.C.C. rings with a cyclic group of units, *Duke Math. J.*, **34** (1967), 243–248; MR **35**: 5467.
2. Artinian rings with a cyclic quasi-regular group, *Duke Math. J.*, **36** (1969), 43–47. See remark following **4.4.6**.

EVERETT, C. J., Jr.
1. An extension theory for rings, *Amer. J. Math.*, **64** (1942), 363–370; MR **4**: 69. Ring analogue of Schreier extension theory for groups.

FARAHAT, H. K.
1. The multiplicative groups of a ring, *Math. Z.*, **87** (1965), 378–384; MR **31**: 209. Given multiplicative subgroups of a ring R, finds those of homomorph of R, eRe, and $M_n(R)$. For finite rings gets orders of subgroups.

FARAHAT, H. K., MIRSKY, L.
1. Group membership in rings of various types, *Math. Z.*, **70** (1958), 231–244; MR **21**: 2676. Study of multiplicative groups in rings, their relations to each other, and criteria for group membership.

FREĬDMAN, P. A.
1. Letter to the editors (concerning a paper of M. Šperling) (Russian), *Mat. Sb.* (*N.S.*), **52 (94)** (1960), 915–916; MR **23**: A168. Correction of Šperling [1].
2. Rings with idealizer condition, I (Russian), *Izv. Vysš. Učebn. Zaved. Matematika* **1960**, No. 2 (15), 213–222; MR **24**: A2598. Begins study of rings in which every proper subring is an ideal in a properly larger subring, called *U*-rings. The class of nil *U*-rings lies strictly between the class of nilpotent rings and the class of locally nilpotent rings.
3. Rings with idealizer condition, II (Russian), *Učen. Zap. Ural. Gos. Univ.*, **1959**, vyp. 23, 35–48; MR **30**: 2038. Finitely generated *U*-rings with minimum condition and *U*-rings generated by a single element.
4. Rings with idealizer condition, III (Russian), *Učen. Zap. Ural. Gos. Univ.*, **1960**, vyp. 23, 49–61; MR **30**: 2039. Finitely generated *U*-rings, *U*-rings with chain conditions. Algebras with idealizer condition.
5. Rings with a distributive lattice of subrings (Russian), *Mat. Sb.* (*N.S.*), **73 (115)** (1967), 513–534; MR **35**: 5474. Characterizes all rings described in the title. For nil *p*-rings these are either cyclic, quasi-cyclic (and thus null), or the *p*-algebra *A*11.

FROBENIUS, G.
1. Theorie der hyperkomplexen Grössen, *Sitz. preuss. Akad. Wiss.*, **1903**, 504–537.
2. Theorie der hyperkomplexen Grössen, II, *Sitz. preuss. Akad. Wiss.*, **1903**, 634–645. Elementary properties of nilpotent algebras. See **1.2.4**.

FUCHS, L.
1. A remark on the Jacobson radical, *Acta. Sci. Math. Szeged*, **14** (1952), 167–168; MR **13**: 903. Frattini subring, as intersection of the maximal right ideals of a ring with identity, is the set of non-generators.
2. *Abelian Groups*, Budapest, 1958; MR **21**: 5672.

GERSTENHABER, M.
1. On nilalgebras and linear varieties of nilpotent matrices, I., *Amer. J. Math.* (2) **80** (1958), 614–622; MR **20**: 3161. Every linear space of nilpotent matrices is conjugate to a space of strictly triangular matrices.
2. On nilalgebras and linear varieties of nilpotent matrices, II., *Duke Math. J.*, **27** (1960), 21–31; MR **22**: 4742. Study of the algebra of endomorphisms induced by right multiplication in a commutative, nil, not necessarily associative algebra.
3. On nilalgebras and linear varieties of nilpotent matrices, III, *Ann. Math.* (2) **70** (1959), 167–205; MR **22**: 4743. Generalization of the embedding theorems of Gerstenhaber [1].
4. On dominance and varieties of commuting matrices, *Ann. of Math.* (2) **73** (1961), 324–348; MR **24**: A1926. Study of a partial order, called "dominance", on the similarity classes of $n \times n$ matrices over a field.
5. On nilalgebras and linear varieties of nilpotent matrices, IV, *Ann. of Math.* (2) **75** (1962), 382–418; MR **30**: 2042. Introduction of a class of nilpotent algebras called "anti-semi-simple" and a study of their important properties.

GHENT, K. S.
1. A note on nilpotent algebras in four units, *Bull. Amer. Math. Soc.*, **40** (1934), 331–338; ZM **9**: 243. Constructs the nilpotent algebras of dimension 4 over a field of characteristic not 2. Correction of Allen [1].

I

GILBERT, G. G.
  1. The multiplicative semigroup of a ring, Ph.D. dissertation, Univ. of Nebraska, 1966; DA **27**: 3594B. What can occur as the multiplicative semigroup of a finite ring? Study of a generalization of cyclic rings.

GILMER, R. W., Jr.
  1. Finite rings having a cyclic multiplicative group of units, *Amer. J. Math.*, **85** (1963), 447–452; MR **27**: 4828.

HALL, M., Jr.
  1. The position of the radical in an algebra, *Trans. Amer. Math. Soc.*, **48** (1940), 391–404; MR **2**: 122. Algebras "bound to the radical". Construction of algebras with a given radical.
  2. *The Theory of Groups*, Macmillan, 1959; MR **21**: 1996.

HALL, M., Jr., SENIOR, J. K.
  1. *The Groups of Order 2^n ($n \leq 6$)*, Macmillan, 1964; MR **29**: 5889.

HALL, P.
  1. A contribution to the theory of groups of prime-power order, *Proc. London Math. Soc.* (2) **36** (1933), 29–95; ZM **7**: 291. See §§(4.1)–(4.3).
  2. The Eulerian functions of a group, *Quart. J. Math.*, *Oxford Ser.*, **7** (1936), 134–151; ZM **14**: 104. Generalization of Anzahl results.
  3. The classification of prime-power groups, *J. Reine Angew. Math.*, **182** (1940), 130–141; MR **2**: 211. See §§ (3.1), (3.5).

HART, R.
  1. A note on algebras of nilpotent matrices, *Arch. Math.* **12** (1961), 324–329; MR **24**: A1930. Generalization of certain results of I. Schur on commutative algebras of matrices.

HASHISAKI, J.
  1. On group algebras of prime power groups, Ph.D. dissertation, Univ. of Illinois, 1953.

HATTORI, A.
  1. Inner endomorphisms of an associative algebra, *J. Math. Soc. Japan*, **6** (1954), 40–44; MR **15**: 929.

HAZLETT, O. C.
  1. On the classification and invariantive characterization of nilpotent algebras, *Amer. J. Math*, **38** (1916), 109–138. Invariant series of subalgebras of a nilpotent algebra A, study of invariants dim (A^k/A^{k+1}), $1 \leq k < \exp A$.

HERSTEIN, I. N.
  1. *Theory of Rings*, Lecture notes, Univ. of Chicago, 1961.
  2. *Topics in Ring Theory*, Lecture notes, Univ. of Chicago, 1965.
  3. *Noncommutative Rings*, Math. Assoc. Amer., Carus Monograph No. 15, 1968.

HIGMAN, G.
  1. Enumerating p-groups, I: inequalities, *Proc. London Math. Soc.* (3), **10** (1960), 24–30; MR **22**: 4779. Lower bounds for the number of groups of order p^n. See **5.2.1**.

2. Enumerating p-groups, II: problems whose solution is PORC, *Proc. London Math. Soc.* (3), **10** (1960), 566–582; MR **23**: A930. A function on the integers is called PORC if there is an integer n such that the function is a polynomial on each of the residue classes mod n.

HINOHARA, Y.
1. A note on annihilator relations, *Nagoya Math. J.*, **17** (1960), 159–160; MR **23**: A913. See (8) of **1.2.1**.

HOLVOET, R.
1. Sur les Z_2 — algèbres du groupe diédral d'ordre 8 et du groupe quaternionique, *C. R. Acad. Sci. Paris (A–B)*, **262** (1966), A209–210; MR **32**: 5726. Shows they are non-isomorphic.

HOPKINS, C.
1. Nilrings with minimal condition for admissible left ideals, *Duke Math. J.*, **4** (1938), 664–667; ZM **20**: 1. Proves such rings are nilpotent.
2. Rings with minimal condition for left ideals, *Ann. of Math.*, **40** (1939), 712–730; MR **1**: 2. See §(3.4).

IVANOVA, O. A.
1. Nilpotent decompositions of associative algebras (Russian), *Mat. Sb. (N.S.)*, **71** (113) (1966), 423–432; MR **34**: 2624.

JACOBSON, N.
1. *The Theory of Rings*, Amer. Math. Soc. Math. Surveys, No. 2, 1943; MR **5**: 31.
2. The radical and semi-simplicity for arbitrary rings, *Amer. J. Math.* **67** (1945), 300–320; MR **7**: 2. Definition of the radical in terms of the circle composition.
3. *Structure of Rings*, Amer. Math. Soc. Colloq. Publ. v. 37, 1956, 2nd ed., 1965; MR **18**: 373, **36**: 5158.

JENNINGS, S. A.
1. The structure of the group ring of a p-group over a modular field, *Trans. Amer. Math. Soc.*, **50** (1941), 175–185; MR **3**: 34. Connection between the powers of the radical of the group algebra and a series of characteristic subgroups of the p-group.
2. A note on chain conditions in nilpotent rings and groups, *Bull. Amer. Math. Soc.*, **50** (1944), 759–763; MR **6**: 114. See **3.4.2**.

JONES, A., SCHÄFFER, J. J.
1. Concerning the structure of certain rings (Spanish), *Bol. Fac. Ingen. Agrimens. Montevideo*, **6** (1957–58), 327–335; MR **21**: 680. Rings in which all additive subgroups are ideals.

KALOUJNINE, L.
1. Zum Problem der Klassifikation der endlichen metabelschen p-Gruppen, *Wiss. Z. Humboldt-Univ. Berlin Math.-Natur. Reihe*, **4** (1954/55), 1–7; MR **17**: 234. See **1.6.7**.

KAPLANSKY, I.
1. *Infinite Abelian Groups*, Univ. of Michigan, Ann Arbor, 1954; MR **16**: 444.

KERTÉSZ, A.
1. A characterization of the Jacobson radical, *Proc. Amer. Math. Soc.*, **14** (1963), 595–597; MR **27**: 1473. The Jacobson radical is the set of non-generators (as right ideals). See remark following **4.1.2**.
2. *Vorlesungen über Artinsche Ringe*, Akadémiai Kiadó, Budapest, 1968.

KOSTRIKIN, A. I., ŠAFAREVIČ, I. R.
1. Groups of homologies of nilpotent algebras (Russian), *Dokl. Akad. Nauk SSSR (N.S.)*, **115** (1957), 1066–1069; MR **19**: 1156.

KÖTHE, G.
1. Die Struktur der Ringe, deren Restklassenring nach dem Radikal vollständig reduzibel ist, *Math. Z.*, **32** (1930), 161–186. See **4.6.2**.

KOVACH, L. D.
1. On elementary nilpotent algebras, Ph.D. dissertation, Purdue University, 1951.

KRUSE, R. L.
1. Rings with periodic additive group in which all subrings are ideals, Ph.D. dissertation, California Institute of Technology, 1964; DA **25**: 4725. See §(4.6).
2. Rings in which all subrings are ideals, I., *Canad. J. Math.*, **20** (1968), 862–871. Reduction of characterization of such rings to the case of finite nilpotent p-rings. See §(4.6).
3. Rings in which all subrings are ideals, II. To appear. Classification of such rings among the finite nilpotent p-rings. See § (4.6).
4. On the adjoint group of a finitely generated radical ring. *J. London Math. Soc.*, (2) **1** (1969). See remark after **1.6.5**.
5. On the circle groups of nilpotent rings. *Amer. Math. Monthly*, **77** (1970). See **1.6.8**.
6. Nilpotent algebras of dimension four. To appear. See Chapter 6.

KRUSE, R. L., PRICE, D. T.
1. On the subring structure of finite nilpotent rings, *Pacific J. Math.*, 1969. See Chapter 4.
2. On the classification of nilpotent rings. To appear. See Chapter 3.
3. Enumerating finite rings. *J. London Math. Soc.* (2), **1** (1969). See Chapter 5.

KUROŠ, A. G.
1. Zur Zerlegung unendlicher Gruppen, *Math. Ann.*, **106** (1932), 107–113; ZM **3**: 243. See **3.4.3**.
2. *Lectures on General Algebra*, Moscow, 1962; English translation: Chelsea, 1963; MR **25**: 5097; **28**: 1228.

LEEUWEN, L. C. A. van
1. Ring extension theory, Dissertation, Technical Institute, Delft, June, 1957; MR **18**: 869.
2. Holomorphe von endlichen Ringen, *Nederl. Akad. Wetensch. Proc. Ser. A*, **68** = *Indag. Math.* **27** (1964), 632–645; MR **33**: 141. Holomorphs of p-rings of small orders.

LEVITZKI, J.
1. On multiplicative systems, *Compositio Math.*, **8** (1950), 76–80; MR **11**: 489. A nil ring with maximal condition is nilpotent.

LEWIN, J.
1. Subrings of finite index in finitely generated rings, *J. Algebra*, **5** (1967), 84–88; MR **34**: 196. See remark after **3.4.7**.

LIEBECK, H.
1. The automorphism group of finite p-groups, J. *Algebra*, **4** (1966), 426–432; MR **34**: 7653. See **4.2.9–4.2.12**.

LIU, S.-X.
1. On algebras in which every subalgebra is an ideal (Chinese), *Acta Math. Sinica*, **14** (1964), 532–537; English translation: *Chinese Math.-Acta* **5** (1964), 571–577; MR **30**: 3115. See remarks after **4.6.10**.

MAHARADZE, L. M.
1. Topological nilpotent rings with minimal condition (Russian), *Uspehi Mat. Nauk (N.S.)* **12**, No. 4 (76) (1957), 181–186; MR **19**: 1063.

MAL'CEV, A. I.
1. On the representation of an algebra as a direct sum of the radical and a semi-simple subalgebra, *C. R. (Doklady) Acad. Sci. URSS (N.S.)*, **36** (1942), 42–45; MR **4**: 130. See **1.6.2** and **4.2.4**
2. Generalized nilpotent algebras and their associated groups (Russian), *Mat. Sb. (N.S.)*, **25** (67) (1949), 347–366; MR **11**: 323. Relations, via the circle composition, of certain classes of locally nilpotent rings and certain infinite p-groups.

MARSHALL, E. I.
1. The Frattini subalgebra of a Lie algebra, *J. London Math. Soc.*, **42** (1967), 416–422.

MASCHKE, H.
1. Über den arithmetischen Charakter der Coefficienten der Substitutionen endlicher linearer Substitutionsgruppen, *Math. Ann.* **50** (1898), 492–498. An ordinary group algebra of a finite group is semi-simple.

MCCOY, N. H.
1. Subdirect sums of rings, *Bull. Amer. Math. Soc.*, **53** (1947), 856–877; MR **9**: 77. Survey paper on subdirect sums.
2. *The Theory of Rings*, MacMillan, 1964; MR **32**: 5680.

MOLIEN, T.
1. Über Systeme höherer complexer Zahlen, *Math. Ann.* **41** (1893), 83–156. Relates algebras, bilinear forms, and algebraic equations.

MORIYA, M.
1. On the automorphisms of a certain class of finite rings, *Kōdai Math. Sem. Rep.*, **18** (1966), 357–367; MR **34**: 5875. Generalization of total nilpotent algebras.

NAGATA, M.
1. *Local Rings*, Interscience, 1962; MR **27**: 5790.

OKUZUMI, M.
1. Automorphisms of a free nilpotent algebra, *Kōdai Math. Sem. Rep.* **20** (1968), 374–384. Decomposition of the automorphism group into "monic" and "diagonal" automorphisms, in a sense similar to **4.2.4–4.2.6**.

O'MEARA, O. T.
1. *Introduction to Quadratic Forms*, Springer Verlag, 1963; MR **27**: 2485.

OUTCALT, D. L.
1. Power-associative algebras in which every subalgebra is an ideal, *Pacific J. Math.*, **20** (1967), 481–485; MR **34**: 7599.

PALMER, D. J.
1. On the structure of nilpotent algebras, Ph.D. dissertation, Univ. of Maryland, 1965; DA **26**: 6747. How does the structure of a nilpotent algebra depend on the structure of its coefficient field?

PASSMAN, D. S.
1. The group algebras of groups of order p^4 over a modular field, *Mich. Math. J.*, **12** (1965), 405–415; MR **32**: 2492. The group algebras are isomorphic if and only if the groups are.

PAVLOV, I. A.
1. On commutative nilpotent matrix algebras (Russian), *Dokl. Akad. Nauk BSSR*, **12** (1968), 393–396; MR **36**: 2650. Conjugate classes of maximal commutative nilpotent subalgebras of exponent $n - 1$ of the algebra of all $n \times n$ matrices over a field.

PEINADO, R. E.
1. On finite rings, *Math. Mag.*, **40** (1967), 83–85; MR **35**: 224. Cyclic rings.

PEIRCE, B.
1. Linear associative algebra, *Amer. J. Math.*, **4** (1881), 97–229. Classic, written 1870. First definition and important properties of abstract associative algebras.

PERLIS, S.
1. A characterization of the radical of an algebra, *Bull. Amer. Math. Soc.*, **48** (1942), 128–132; MR **3**: 264. Definition of the circle composition. See **1.6.2**.

PICKERT, G.
1. Neue Methoden in der Strukturtheorie der kommutativ-assoziativen Algebren, *Math. Ann.*, **116** (1938), 217–280; ZM **22**: 196. Constructs the commutative algebras of small dimensions.

RÉDEI, L.
1. Die Vollidealringe, *Monatsh. Math.*, **56** (1952), 89–95; MR **14**: 127. Rings in which all additive subgroups are ideals.
2. Vollidealringe im weiteren Sinn, I, *Acta Math. Acad. Sci. Hungar.*, **3** (1952), 243–268; MR **14**: 941. Characterizes rings generated by one element, in which every subring is an ideal.
3. Die Holomorphentheorie für Gruppen und Ringe, *Acta Math. Acad. Sci. Hungar.*, **5** (1954), 169–195; MR **17**: 342.
4. *Algebra, Erster Teil*, Leibzig, 1959; MR **21**: 4885.

SCHEUNEMAN, J.
1. Two-step nilpotent Lie algebras, *J. Algebra*, **7** (1967), 152–159. Study of nilpotent Lie algebras of exponent 3 by means of a certain duality and an arithmetic isomorphism invariant.

SCORZA, G.
1. *Corpi Numerici e Algebre*, Messina, Cosa Editrice Giuseppe Principato, 1921.

2. Le algebre doppie, *Rend. R. Acad. Sci. Fis. Mat. Napoli*, (3) **28** (1922), 65–79. Algebras of dimension 2, with automorphism groups.
3. Sulla struttura delle algebre pseudonulle, *Rend. Accad. Lincei*, (6) **20** (1934), 143–149; ZM **10**: 195.
4. Le algebre per oguna delle quali la sotto-algebra eccezionale è potenziale, *Atti Accad. Sci. Torino Cl. Sci. Fis. Mat. Natur.*, **70** (1934–35), 26–45; ZM **11**: 199. Finite-dimensional algebras whose radical is a power algebra.
5. Sopra una classe di algebre pseudonulle, *Atti Accad. Sci. Torino Cl. Sci. Fis. Mat. Natur.*, **70** (1934–35), 196–211; ZM **12**: 7. Characterization of the nilpotent algebras of dimension n and exponent n.
6. Le algebre del 3° ordine, *Mem. Real. Accad. Sci. Fis. Mat. Napoli*, (2) **20** (1935), n. 13; ZM **12**: 392. Determination, by generators and relations, of all algebras of dimension 3. See **2.3.6**.
7. Le algebre del 4° ordine, *Mem. Real. Accad. Sci. Fis. Mat. Napoli*, (2) **20** (1935), n. 14, and *Giorn. Mat. Battaglini* (3) **73** (1935), 129–212; ZM **12**: 392. Determination, by generators and relations, of all algebras of dimension 4. Not always reduced to mutually non-isomorphic cases. See Chapter 6.
8. Sulle algebre pseudonulle di ordine massimo, *Ann. Mat. Pura Appl.*, (4), **14** (1936), 1–14; ZM **12**: 102. Free nilpotent algebras.
9. *Opere Scelte*, 3 vols., Rome: Edizione Cremonese, 1962.

SEHGAL, S. K.
1. On the isomorphism of group algebras, *Math. Z.*, **95** (1967), 71–75; MR **34**: 5950. Study of modular group algebras of a class of p-groups which contains the class of regular p-groups.

SHAW, J. B.
1. Theory of linear associative algebra, *Trans. Amer. Math. Soc.*, **4** (1903), 251–287. Matrix representation, etc.
2. On nilpotent algebras, *Trans. Amer. Math. Soc.*, **4** (1903), 405–422. Representation in strictly lower triangular matrices. Structure of nilpotent algebras which are "nearly" power algebras.

SHODA, K.
1. Über die Automorphismen einer endlichen Abelschen Gruppe, *Math. Ann.*, **100** (1928), 674–686. Endomorphism ring of a finite abelian p-group.
2. Über die Einheitengruppe eines endlichen Ringes, *Math. Ann.*, **102** (1930), 273–282. Primary decomposition with the term "p-ring". See §(1.5).

SIMS, C.
1. Enumerating p-groups, *Proc. London Math. Soc.* (3) **15** (1965), 151–166; MR **30**: 164. Asymptotic results. See §(5.2).

SMITH, G. W.
1. Nilpotent algebras generated by two units, i and j, such that i^2 is not an independent unit, *Amer. J. Math.*, **41** (1919), 143–164.

ŠPERLING, M.
1. On rings, every subring of which is an ideal (Russian), *Mat. Sb. (N.S.)*, **17 (59)** (1945), 371–384; MR **7**: 509. Incomplete determination of rings of title which are generated by one element.

SUPRUNENKO, D. A.
1. On maximal commutative subalgebras of the full linear algebra (Russian), *Uspehi Mat. Nauk.* (*N.S.*), **11** (1956) No. 3 (69), 181–184; MR **18**: 639. There are an infinite number of conjugate classes of maximal commutative subalgebras of the algebra of all $n \times n$ matrices over an algebraically closed field of characteristic 0, provided $n > 6$. [If $n \leq 6$ the number is finite.]
2. Maximal commutative nilpotent subalgebras of class $n\text{-}2$ of a complete matrix algebra (Russian), *Veci Akad. Navuk BSSR, Ser. Fiz.-Tehn. Navuk*, **1956**, No. 3, 135–145.

SUPRUNENKO, D. A., TYŠKEVIČ, R. I.
1. *Commutative Matrices* (Russian), Nauka i Tehnika, Minsk, 1966; English translation, Academic Press, 1968; MR **34**: 1356.

SZÁSZ, F.
1. On rings every subring of which is a multiple of the ring, *Publ. Math. Debrecen*, **4** (1956), 237–238; MR **18**: 187.
2. On rings such that every subring is a direct summand of the ring (Russian), *Mat. Sb.* (*N.S.*) **40** (**82**) (1956), 269–272; MR **18**: 788.
3. Über die homomorphen Bilder des Ringes der ganzen Zahlen und über eine verwandte Ringfamilie, *Monatsh. Math.*, **61** (1957), 37–41; MR **19**: 9.
4. Ringe, deren echte Unterringe streng zyklische Rechtsideale sind, *Magyar Tud. Akad. Mat. Kutató Int. Közl.*, **5** (1960), 287–292; MR **27**: 3660.
5. Die Ringe, deren endlich erzeugbare echte Unterringe Hauptrechtsideale sind, *Acta Math. Acad. Sci. Hungar.*, **13** (1962), 115–132; MR **25**: 3062.

SZELE, T.
1. Ein Satz über die Struktur der endlichen Ringe, *Acta Sci. Math.* (*Szeged*), **11** (1948), 246–250; MR **10**: 96. Matrix representation. See **2.1.1**.
2. Zur Theorie der Zeroringe, *Math. Ann.*, **121** (1949), 242–246; MR **11**: 496. What abelian groups can occur as additive groups only for null rings?
3. Nilpotent Artinian rings, *Publ. Math. Debrecen*, **4** (1955), 71–78; MR **17**: 122. See §(3.4).

TOSKEY, B. R.
1. Rings on a direct sum of cyclic groups, *Publ. Math. Debrecen*, **10** (1963), 93–95; MR **30**: 117. Formulation, in terms of matrices and Kronecker products, of the isomorphism problem for rings with the same additive group.
2. A system of canonical forms for rings on a direct sum of two infinite cyclic groups, *Pacific J. Math.* **20** (1967), 179–188; MR **34**: 4303.

VANDIVER, H. S.
1. Theory of finite algebras, *Trans. Amer. Math. Soc.*, **13** (1912), 293–304. Studies units. Finite commutative algebras with at least one non-divisor of 0.

WAERDEN, B. L. van der
1. *Algebra I, II.*, Springer Verlag.

WARD, H. N.
1. Some results on the group algebra of a group over a prime field, mimeographed notes, *Seminar on Finite Groups and Related Topics*, Harvard University, 1960–1961, 13–19.

WATTERS, J. F.
1. On the adjoint group of a radical ring, *J. London Math. Soc.* **43** (1968), 725–729. See remark following **1.6.5**.

WEDDERBURN, J. H. M.
1. A theorem on finite algebras, *Trans. Amer. Math. Soc.*, **6** (1905), 349–352. Finite division algebras are commutative.
2. On hypercomplex numbers, *Proc. London Math. Soc.* (2), **6** (1908), 77–118. The classic structure theory for semi-simple algebras.
3. *Lectures on Matrices*, Amer. Math. Soc. Colloq. Publ., **17**, 1934; ZM **10**: 99.
4. Note on algebras, *Ann. Math.*, **38** (1937), 854–856; ZM **18**: 103. See **2.4.2**.

WEISNER, L.
1. Some properties of prime-power groups, *Trans. Amer. Math. Soc.* **38** (1935), 485–492; ZM **12**: 394. Anzahl results.

WIEGANDT, R.
1. Über transfinit nilpotente Ringe, *Acta Math. Acad. Sci. Hungar.*, **17** (1966), 101–114; MR **34**: 210. Extends Szele [3] to "transfinite nilpotent" rings and inverse "system" of finite nilpotent rings.

List of special notation

a.c.c	ascending chain condition	42
Aut $(R;I)$	group of automorphisms of a ring R which fix R/I elementwise, I an ideal of R	56
$\mathfrak{A}_l(S)$	left annihilator of a subset S of a ring	2
$\mathfrak{A}_r(S)$	right annihilator of S	2
$\mathfrak{A}(S)$	two-sided annihilator of S	2
$C(p^n)$	cyclic group of order p^n	1
$C(p^\omega)$	Prüfer quasi-cyclic p-group	1
char	characteristic	1
d.c.c	descending chain condition	40
dim	dimension of an algebra or vector space	
exp	exponent	1
F^*	multiplicative group of the field F	2
$F[S]$	free ring on a set S (§2.2 only)	19
$FN_e[S]$	free nilpotent ring of exponent e on S	20
$F[G]$	group algebra of G over a field F (§2.4 only)	24
$GF(q)$	finite field of q elements	2
$GL(n, q)$	general linear group in n dimensions over $GF(q)$	82
$_p[n_1, \ldots, n_k]$	type of an abelian p-group	1
R_p	(1) set of elements $x \in R$ with $px = 0$	2
	(2) the p-component of a torsion ring R	9
R^+	additive group of a ring R	1
$\lvert R \rvert$	order of a finite ring R	1
$[R:S]$	index of a subring S in a ring R	1
$U[n, s]$	§4.4 only: class of nilpotent rings of order p^n with a unique subring of order p^s	65
$x \circ y$	circle composition	11
$X\, n_1 \ldots n_k\, y$	family notation for finite nilpotent p-rings	34
$0[n^\alpha]$	big 0 notation	80–81
Φ_R	Frattini subring of a ring R	51
$\cong$	isomorphism	1
$\overset{\cong}{F}$	family equivalence relation	27
$S + T$	ring sum of subrings S and T	1
$S - T$	set difference	1

$\oplus$ ring direct sum   1
$\dotplus$ §4.6 only: additive group direct sum   69
                        §5.3 only: module direct sum   92
$\otimes$ Kronecker (tensor) product
$\langle S \rangle$ subring generated by S 1
$\{S\}$ (1) [unordered] set
                        (2) ordered basis
                        (3) §4.6 only: additive subgroup spanned by S 69
$[S]_r$ §4.1 only: right ideal generated by S 51

adjoint group 10–16
adjunction of identity 10
algebras 2
  in the same family 31
  in which all subalgebras are
    ideals 73
  power 22
Allen, R. B. 99
almost annihilates, almost-null 71
analogy with groups 4–5
Andrijanov, V. I. 79
annihilator 2
  series 4
Anzahl results 59–62
associativity vi, 2
  in machine computation 18
asymptotic counting 80–94
  results for nilpotent rings 80–90
  results for non-nilpotent
    rings 90–94
automorphisms 54–59
  diagonal 55
  inner 55
  monic 55
  of free nilpotent algebras 56, 58
  of total nilpotent algebras 55–56
  outer 55

Ballieu, R. 98
basis 1
basis theorem 53
Baumgartner, K. 17
Beaumont, R. A. 17
Bechtell, H. 53
Boyce, F. W. 99, 102
Burnside basis theorem 53

capability 42
  of direct sum 48–49
  of nilpotent rings R with
    $|R^2| = p$ 49–50
  of null rings 44–45

of subdirect sum 48
of total nilpotent algebras 45
capable ideal 44
characteristic 1
Chinese remainder theorem 10
circle composition 11
  group 10–16
  group cyclic 65–66
class of
  automorphism group 56–59
  circle group 13–16
Coleman, D. G. 25–26
computational techniques 18, 20–21,
  34
conditions for capability 48–50
construction of families 42–44
  of nilpotent rings 20
  of nilpotent rings of order p^3 34–40
  of p-rings from p-algebras 32–34
counting finite rings 80–94
cyclic 1
  rings 21–22

degree (of total nilpotent algebra) 18
Deskins, W. E. 25–26
diagonal automorphisms 55
Dickson, L. E. 24, 69
Dubisch, R. 18–19, 55–56

Eggert, N. H. 13
Eldridge, K. E. 13, 65–66
enumeration results 59–62
exponent 1

family classification 27
  notation for finite nilpotent
    p-rings 34
  of algebras 31
  of nilpotent rings of orders p, p^2 34
  of nilpotent rings of order p^3 35–40
  of rings R with $pR^2 = 0$ 32
Fischer, I. 13, 65–66

Frattini subring　51
free nilpotent algebra　20
　automorphisms of　56, 58
free nilpotent ring　20
free ring　19–20
Freĭdman, P. A.　70, 79
Frobenius, G.　3, 22
Fuchs, L.　51–53

generating rings　51–53
generators and relations　20–24, 95–108
Ghent, K. S.　99
Gilmer, R. W.　13
group algebra　24
group analogy　4–5

H-algebra　73
Hall, M.　91
Hall, P.　27–31, 42–44, 54, 59
Hamiltonian rings　68–79
height　1
Higman, G.　81
Hinohara, Y.　3
Holvoet, R.　25
Hopkins, C.　41
H-ring, H-p-ring　68

ideal　2
idempotents　69
identity, adjunction of　10
indeterminate　19
index　1
inner automorphism　55
invertible　10

Jacobson, N.　52–53
Jacobson radical ring　11
Jennings, S. A.　40–41
Jones, A.　79

Kaloujnine, L.　13–14
Kertész, A.　53
Köthe, G.　69
Kronecker product　17, 25
Kulakoff's theorem　62
Kuroš, A.　40–41

left annihilator　2
　series　19

Lewin, J.　42
Lie algebra　vi
Liebeck, H.　56–58
Lifting idempotents　69
Liu, S.-X.　73
lower annihilator series　4

Maclagan-Wedderburn, J. H.—see
　Wedderburn, J. H. M.
Mal'cev, A. I.　11, 55
Marshall, E.　53
Maschke, H.　24
matrix representation　17, 88–89
maximal family　44
　ideal　6
　right ideal　51–52
　subring　6, 53
McCoy, N.　7, 9
modular group algebras of
　p-groups　24
monic automorphism　55
monomial　19
Montgomery, D.　9
Moriya, M.　56

nil　1
nilpotent　1
　algebras of dimension 3　23
　algebras of dimension 4　99–108
　Artinian rings　40, 52
　Noetherian rings　42
　rings of order p^2　22
　rings of order p^3　34–40
　rings of type $_p(n, 1)$　96–98
　rings with ascending chain
　　condition　42
　rings with descending chain
　　condition　40–42, 52
　rings with only one subring of a
　　given order　62–68
non-generator　51
notation for finite nilpotent p-rings　34
null　1
　rings, capability of　44–45

Okuzumi, M.　56
order　1
Outcalt, D.　73
outer automorphism　55

p-algebra 2
Passman, D. S. 25
Peirce, B. 70
  decomposition 70
Perlis, S. 11, 18–19, 55–56
power algebra 21–22
power rings 21, 53
  of type $_p(2, 1, 1)$ 22
  of type $_p(2, 2)$ 95
  of type $_p(n, 1)$ 96
primary decomposition 9
p-ring 9

q-algebra 2
quasi-cyclic 1
quasi-regular 11

radical ring 11
rank 1
Rédei, L. 79
refinement theorem 6
right annihilator 2
  annihilator series 19
  Frattini subring 51
  non-generator 51
rings, cyclic 21–22
  in a family 29–34
  in the same family as an algebra 32
  in which all subrings are
    ideals 68–79
  of matrices 17

Schäffer, J. J. 79
Schuind, M.-J. 98

Schur multiplier 42
Scorza, G. 20, 99
Shaw, J. B. 18
Shoda, K. 9
Sims, C. 5–6, 85–88
span 1
Sperling, M. 79
stem ring 30
structure constants 21
subdirect sum 7
  of capable rings 48
subdirectly irreducible 7
subring generated by S 1
Szele, T. 18, 40–41

tensor product 17, 25
torsion, torsion-free 9
Toskey, B. R. 17
total nilpotent algebras 18
  automorphisms of 55–56
  capability of 45
two-sided annihilator 2
type 1
type I, II element in an H-p-ring 74

unique subrings 62–68
upper annihilator series 4

Watters, J. F. 12
Wedderburn, J. H. M. 25, 91
  structure theory, summary 91
Wiegandt, R. 41

Zorn's Lemma 52
0-notation 80–81